Key Questions in Ecc
A Study and Revision Guide

Key Questions in Ecology: A Study and Revision Guide

Paul A. Rees, *BSc (Hons), LLM, PhD, CertEd*

Senior Lecturer
School of Science, Engineering and Environment,
University of Salford, UK

CABI is a trading name of CAB International

CABI	CABI
Nosworthy Way	WeWork
Wallingford	One Lincoln St
Oxfordshire OX10 8DE	24th Floor
UK	Boston, MA 02111
	USA

Tel: +44 (0)1491 832111
Fax: +44 (0)1491 833508
E-mail: info@cabi.org
Website: www.cabi.org

T: +1 (617)682-9015
E-mail: cabi-nao@cabi.org

A catalogue record for this book is available from the British Library, London, UK.

Library of Congress Cataloging-in-Publication Data

Names: Rees, Paul A., author.
Title: Key questions in ecology : a study and revision guide / Paul A. Rees, BSc (Hons), LLM, PhD, CertEd, Senior Lecturer, School of Science, Engineering and Environment, University of Salford, UK.
Description: Wallingford, Oxfordshire ; Boston, MA : CAB International, [2021] | Series: Key questions | Includes bibliographical references and index. | Summary: "This book is intended as a study and revision guide for students following programmes of study in which ecology is an important component. It contains 500 multiple-choice questions (and answers) set at three levels - foundation, intermediate and advanced"-- Provided by publisher.
Identifiers: LCCN 2020031736 (print) | LCCN 2020031737 (ebook) | ISBN 9781789247572 (paperback) | ISBN 9781789247589 (ebook) | ISBN 9781789247596 (epub)
Subjects: LCSH: Ecology.
Classification: LCC QH541 .R355 2021 (print) | LCC QH541 (ebook) | DDC 577--dc23
LC record available at https://lccn.loc.gov/2020031736
LC ebook record available at https://lccn.loc.gov/2020031737

References to Internet websites (URLs) were accurate at the time of writing.

ISBN-13: 978 1 78924 757 2 (paperback)
        978 1 78924 758 9 (ePDF)
        978 1 78924 759 6 (ePub)

Commissioning Editor: Ward Cooper
Editorial Assistant: Emma McCann
Production Editor: James Bishop

Typeset by SPi, Pondicherry, India
Printed and bound in the UK by CPI Group, (UK) Ltd, Croydon, CR0 4YY

Contents

About the Author

Paul Rees is a Senior Lecturer in the School of Science, Engineering and Environment at the University of Salford, UK. He holds a PhD in animal ecology and behaviour from the University of Bradford. Paul has previously lectured at three Further Education Colleges and a Higher Education College in the UK, and trained biology teachers at Sokoto College of Education in Nigeria. He has taught from GCE 'O'/GCSE level to MSc level and has been an external examiner for a range of taught programmes, from Higher National Diploma to MSc level, at six British universities. Paul has published papers on mammal behaviour and ecology, wildlife law, and the role of zoos in conservation, along with seven textbooks concerned with ecology, zoo biology and wildlife law.

Preface

Since I began studying ecology in the early 1970s the subject has expanded beyond recognition. The textbook I used to study zoology in the sixth form (Grove and Newell's *Animal Biology*) contained just 17 pages on ecology in a volume of 911 pages (less than 2% of the book). Now, ecology is a major scientific discipline in its own right and the academic and public interest in the subject is vast.

Everyone should know some basic ecology. Our very existence depends upon this. The more ecologists we train, regardless of whether or not they ever find employment as professional ecologists, the safer the planet will be. We need more teachers, politicians and ordinary members of the public to understand ecology, whether they live in biologically diverse rural areas or large cities.

A old colleague of mine had a sign on the door of his classroom that said 'Geography is everywhere, and geology is holding it up'. This is self-evidently true. It is also true that ecology is everywhere. Just how long that remains the case depends upon us, our children and their children caring enough to protect all the systems that make ecology work. I hope some people who are not, in any formal sense, students of ecology will find this book a useful introduction to some of the history, development and key concepts of ecology. However, it is primarily intended as a study and revision guide for students who have already begun learning about the subject and I hope that it will encourage them to find out more for themselves.

Acknowledgements

My thanks go to Ward Cooper (Commissioning Editor at CABI) for enthusiastically supporting this project and for coming up with the title. I am also grateful to my daughter Clara who very kindly checked the manuscript for errors without payment, although having the opportunity to correct her father should have been payment enough. Any errors that remain are mine. This book was written and produced during the Covid-19 pandemic by people who were working under very difficult conditions and I thank them all for their efforts.

How to Use This Book

The questions are arranged by topic and divided into three levels: foundation, intermediate and advanced. These levels are not intended to reflect any particular curriculum but rather general levels of difficulty, and should not be taken too seriously. Knowledge of definitions and basic facts are dealt with at the foundation level; the intermediate level contains questions on methods, processes and basic calculations; and the advanced level contains questions involving more obscure facts and definitions, along with calculations, and mathematical concepts where appropriate. However, there is some variation between chapters as not all of the topics covered lend themselves to this approach. Students are advised to check the syllabuses they are following in detail before relying too much on this book as a preparation for specific exams.

Students are encouraged to complete a whole chapter – or at least a complete section (foundation, intermediate or advanced) – before looking at the answers. This is because the explanations for some answers may assist in selecting the correct answer to subsequent questions, although I have tried to avoid this where possible. The order in which the chapters are attempted does not really matter because each is about a different area of ecology. However, within any chapter you are advised to attempt the foundation questions first, followed by the intermediate questions and finally the advanced questions. Some of the questions involve calculations and it would be useful to have access to a calculator when attempting them.

1

The History and Foundations of Ecology

This chapter contains questions about the history of ecology, historically important scientists who have made significant contributions to the science, along with some basic terminology and founding principles.

Foundation

1.1f **The first person to use the term 'ecology' was**

    a. Eugene Odum

    b. Ernst Haeckel

    c. Charles Elton

    d. Arthur Tansley

1.2f **The term 'ecology' was first used in which decade?**

    a. 1840s

    b. 1850s

    c. 1860s

    d. 1870s

1.3f **The term 'ecology' is derived from the Greek word 'oikos' which means**

    a. nature

    b. environment

    c. interrelationships

    d. home

1.4f **'The complex of a biological community and its environment' is one definition of an**

a. ecosystem

b. ecotone

c. ecotype

d. ecological succession

1.5f **The ecology of a single species is known as**

a. uniecology

b. autecology

c. monoecology

d. synecology

1.6f **The person credited with coining the term 'ecosystem' is**

a. Eugene Odum

b. Arthur Tansley

c. Charles Elton

d. Charles Darwin

1.7f **Which of the following statements about an ecosystem is false?**

a. It may be of any size

b. Energy flows between its components

c. Nutrients cycle between its components

d. It must have been formed naturally

1.8f **An anthropogenic factor affecting an ecosystem is related to**

a. human activity

b. soil conditions

c. animal activity

d. weather conditions

1.9f **Which of the following sequences correctly illustrates the levels of organisation in an ecosystem?**

a. organism > deme > population > species > community

b. deme > population > community > species > organism

   c. species > population > deme > organism > community

   d. organism > population > deme > species > community

1.10f **The Rev. Gilbert White, an English clergyman, published an important early account of the natural history of part of England entitled *The Natural History and Antiquities of***

   a. *Selby*

   b. *Surrey*

   c. *Selborne*

   d. *Sussex*

1.11f **The American forester, ecologist and conservationist Aldo Leopold published an influential book entitled**

   a. *The Forest Ecology of California*

   b. *A Sand County Almanac*

   c. *A Natural History of Oregon*

   d. *Diary of a Desert State*

1.12f **Biological activity on Earth is largely restricted to which of the following ranges of temperature?**

   a. $-10°C$–$60°C$

   b. $0°C$–$50°C$

   c. $10°C$–$40°C$

   d. $0°C$–$70°C$

1.13f **The study of the origin, distribution, adaptation and associations of plants and animals is called**

   a. phytogeography

   b. palaeogeography

   c. zoogeography

   d. biogeography

1.14f **An Eltonian pyramid is a concept in the study of**

   a. population ecology

   b. energy flow in ecosystems

c. nutrient cycling

d. interspecies competition

1.15f Who was especially important in the development of botanical geography?

a. Frederic Clements

b. Alfred Russel Wallace

c. Arthur Tansley

d. Alexander von Humboldt

1.16f The science of wildlife management was founded in 1933 when the book *Game Management* was published. It was written by

a. Aldo Leopold

b. Eugene Odum

c. Daniel Janzen

d. Keith Eltringham

1.17f The sustainable use and management of natural resources is referred to as

a. preservation

b. conservation

c. restoration

d. protection

1.18f The term 'flora' means

a. all of the plant life in a particular area at a given time

b. a list of the plant life in a particular area at a given time

c. a field guide to the plant life in a particular area at a given time

d. all of the above

1.19f A sward is an area of ground covered with

a. trees

b. bare soil

c. short grass

d. leaf litter

1.20f **A growth of plants, especially trees or crops, in a particular localised area is called a**

a. frame

b. rank

c. stand

d. stage

Intermediate

1.1i *An Essay on the Principle of Population as it Affects the Future Improvement of Society* **was written by**

a. Thomas Malthus

b. Ernst Haeckel

c. Eugene Odum

d. William Hamilton

1.2i **What is the name of the layer of the atmosphere nearest to the Earth's surface?**

a. The stratosphere

b. The mesosphere

c. The thermosphere

d. The troposphere

1.3i **The surface layer of the ocean is called the**

a. abyssopelagic zone

b. mesopelagic zone

c. epipelagic zone

d. bathypelagic

1.4i **The hydrosphere is**

a. an alternative name for the oceans

b. a type of aquatic plant

c. all of the water on, under and above the Earth's surface

d. all of the water held in the Earth's glaciers and rivers

1.5i   The adjective 'littoral' describes something that is situated

a. on or near the shore of a sea or lake

b. at the edge of a hot desert

c. at the edge of a woodland or forest

d. at the bottom of the ocean

1.6i   Limnology is the study of

a. rock strata

b. inland waters

c. coastal waters

d. mountainous areas

1.7i   Which of the following terms was previously used to mean the same as the modern term 'ecosystem'?

a. Microcosm

b. Biocoenosis

c. Biogeocoenosis

d. All of the above

1.8i   Swimming organisms that are able to navigate at will are collectively referred to as

a. nekton

b. plankton

c. periphyton

d. benthos

1.9i   The contention that the living and non-living components of the Earth form a complex interacting system that can be treated as a single organism is known as the

a. Maia hypothesis

b. Athena hypothesis

c. Gaia hypothesis

d. Rhea hypothesis

1.10i A number of long-term ecological studies have been conducted in a famous woodland near Oxford, England, called

a. Wytham Wood

b. Willow Wood

c. William's Wood

d. Wilton Wood

1.11i Silver Springs is a freshwater ecosystem in Florida that is
historically important because ecologists conducted work
there that has contributed to our understanding of

a. the dynamics of freshwater fish populations

b. energy flow and trophic structure in freshwater ecosystems

c. the effect of deforestation on nutrient quality in rivers

d. eutrophication caused by agricultural ecosystems

1.12i Which part of Table 1.1 accurately pairs each zoologist with
a species which they have studied extensively?

a. A

b. B

c. C

d. D

Table 1.1

A	
George Schaller	Grey wolves
L. David Mech	African lions
Jane Goodall	Chimpanzees
Raman Sukumar	Asian elephants

B	
George Schaller	Asian elephants
L. David Mech	African lions
Jane Goodall	Chimpanzees
Raman Sukumar	Grey wolves

C	
George Schaller	African lions
L. David Mech	Grey wolves
Jane Goodall	Chimpanzees
Raman Sukumar	Asian elephants

D	
George Schaller	African lions
L. David Mech	Chimpanzees
Jane Goodall	Asian elephants
Raman Sukumar	Grey wolves

1.13i **During the early development of ecology as a science, plant and animal ecology focussed on different areas of study. Which of the following statements is true?**

    a. Animal ecologists focussed on population ecology and plant ecologists focussed on community ecology

    b. Animal ecologists focussed on behavioural ecology and plant ecologists focussed on population ecology

    c. Animal ecologists focussed on community ecology and plant ecologists focussed on population ecology

    d. Animal ecologists focussed on population ecology and plant ecologists focussed on production ecology

1.14i **Population studies of plants began later than animal population ecology. Which of the following is unlikely to have been a reason for this?**

    a. Many plants are large and very long-lived

    b. Plants reproduce dormant seeds

    c. Some species reproduce vegetatively and this makes individual plants difficult to define

    d. Most plants are stationary

1.15i **Which of the following countries does not contain wild-living Old World monkeys?**

    a. Brazil

    b. Sri Lanka

c. Malawi

d. Morroco

1.16i **Which of the following species is not a specialist feeder?**

a. The giant panda (*Ailuropoda melanoleuca*)

b. The koala (*Phascolarctos cinereus*)

c. The osprey (*Pandion haliaetus*)

d. The Asian elephant (*Elephas maximus*)

1.17i **Which of the following ecological issues did Rachel Carson bring to the attention of the general public in the 1960s?**

a. The pollution of the oceans with plastic

b. The appearance of a 'hole' in the ozone layer

c. The effect of global climate change on biodiversity

d. The accumulation of pesticides and other chemicals in food chains

1.18i **In the early and mid-1900s two 'competing' groups of botanists appeared in different parts of the world interested in different aspects of plant ecology. The first group were interested in the composition, structure and distribution of plant communities (group 1) and the second were interested in the development of plant communities by the process of succession (group 2). These groups were in**

a. Australia (group1); Europe (group 2)

b. Europe (group 1); United States (group 2)

c. United States (group 1); Australia (group 2)

d. United States (group 1); Europe (group 2)

1.19i **Who published a book entitled *Animal Ecology* in 1927?**

a. Charles Elton

b. Arthur Cain

c. Arthur Tansley

d. Eugene Odum

1.20i Which of the following lists describes the components of the environment originally encompassed by the term 'natural history'?

   a. Animals and plants

   b. Rocks, soils, climate and living things

   c. Living things and rocks

   d. Living things, rocks and soils

Advanced

1.1a **Fossorial animal species are characterised by**

   a. a burrowing habit

   b. an aquatic habit

   c. a tree-climbing habit

   d. a running habit

1.2a **Which of the following statements about zoogeographical realms is false?**

   a. The Nearctic realm consists of North America and Greenland

   b. The Neotropical realm extends from central Mexico to the southern tip of South America

   c. The Oriental realm consists of Southeast Asia, Sumatra, Java, Borneo and the Philippines

   d. The Ethiopian realm consists of all of Africa and southern Arabia

1.3a **Select the part of Table 1.2 which most accurately matches ecological studies with the branch of ecology within which they fall.**

   a. A

   b. B

   c. C

   d. D

Table 1.2

A	
The relationship between mating systems in lions and changes in numbers	Theoretical ecology
Patterns of energy flow in a lake	Production ecology
Historical changes in the combination of plant species occupying derelict agricultural land	Community ecology
A mathematical analysis of species diversity patterns	Behavioural ecology

B	
The relationship between mating systems in lions and changes in numbers	Behavioural ecology
Patterns of energy flow in a lake	Production ecology
Historical changes in the combination of plant species occupying derelict agricultural land	Community ecology
A mathematical analysis of species diversity patterns	Theoretical ecology

C	
The relationship between mating systems in lions and changes in numbers	Behavioural ecology
Patterns of energy flow in a lake	Community ecology
Historical changes in the combination of plant species occupying derelict agricultural land	Production ecology
A mathematical analysis of species diversity patterns	Theoretical ecology

D	
The relationship between mating systems in lions and changes in numbers	Behavioural ecology
Patterns of energy flow in a lake	Theoretical ecology
Historical changes in the combination of plant species occupying derelict agricultural land	Production ecology
A mathematical analysis of species diversity patterns	Community ecology

1.4a The movement of land masses over the Earth's surface during geological time has affected the geographical distribution of organisms. The name of the single land mass that eventually split around 180 million years ago to form present-day Africa, Arabia, Madagascar, India, South America, Australia and Antarctica was

   a. Laurasia

   b. Pangea

   c. Gondwana

   d. Tethys

1.5a The world was first divided into zoogeographical regions by Sclater in 1857 on the basis of its

   a. bird fauna

   b. mammal fauna

   c. flora

   d. reptile fauna

1.6a In which zoogeographical realm do guanacos, rheas and armadillos naturally occur?

   a. Ethiopian

   b. Oriental

   c. Palaearctic

   d. Neotropical

1.7a In which zoogeographical realms do elephants naturally occur?

   a. Neotropical and Ethiopian

   b. Ethiopian and Oriental

   c. Oriental and Neotropical

   d. Palaearctic and Oriental

1.8a F. E. Clements is famous for his work on

   a. plant succession

   b. bird ecology

c. lizard diversity

d. forest ecology

1.9a **The book entitled *The Distribution and Abundance of Animals* was important in establishing ecology as a science and was written by**

a. Pearl and Reed

b. Andrewartha and Birch

c. MacArthur and Wilson

d. Borman and Likens

1.10a **Which of the following is famous for suggesting that the size of animal populations could be regulated by their social behaviour?**

a. Vero Wynne-Edwards

b. Robert MacArthur

c. Vito Volterra

d. Raymond Pearl

1.11a **Which of the following was important in the establishment of the scientific study of plant population dynamics?**

a. A. Macfadyen

b. C. J. Krebs

c. T. R. E. Southwood

d. J. L. Harper

1.12a **The science that studies the historical interconnectedness of nature and human culture is called**

a. chronological ecology

b. historical ecology

c. cultural ecology

d. human ecology

1.13a **Animals that are adapted for climbing are referred to as**

a. scansorial

b. cursorial

c. volant

d. arboreal

1.14a **Who said "The 'balance of nature' does not exist, and perhaps never has existed"?**

a. Charles Darwin

b. Warder Clyde Allee

c. Thomas Park

d. Charles Elton

1.15a **The Bureau of Animal Population was founded in 1932 at the University of**

a. Cambridge

b. Oxford

c. Chicago

d. California

1.16a **Which of the following countries was not visited by Charles Darwin on his voyage on HMS *Beagle*?**

a. New Zealand

b. Australia

c. India

d. Peru

1.17a **In 1858, who read two papers describing a theory of evolution at the Linnean Society in London?**

a. Charles Darwin read both

b. Alfred Russel Wallace read both

c. Charles Darwin read one and Alfred Russell Wallace read the other

d. The Secretary of the Linnean Society, J. J. Bennet, read both

1.18a **An alien species of animal or plant that has become established in the wild in a country where it does not naturally occur, and is breeding and holding its own with native fauna and flora is referred to as being**

a. naturalised

b. feral

c. imported

d. exotic

1.19a **The natural history collection of the British Museum in London was founded on the collection of**

a. Lord Walter Rothschild

b. Sir Richard Owen

c. Sir Hans Sloane

d. Sir Stamford Raffles

1.20a **The belief that plant communities can be identified and classified as discrete units like species and genera is held by ecologists who are exponents of**

a. social botany

b. sociophytology

c. ethology

d. phytosociology

2

Abiotic Factors and Environmental Monitoring

This chapter contains questions about the physical factors in the environment and their measurement.

Foundation

2.1f The term 'edaphic factor' relates to conditions in the

    a. leaf litter

    b. air

    c. water

    d. soil

2.2f A small white container that is suspended above the ground on legs and designed to shelter weather recording equipment is called a

    a. Stevenson screen

    b. Petersen screen

    c. Andersen screen

    d. Gunderson screen

2.3f Which of the following terms is not used to describe the highest altitude at which trees can survive on a mountainside?

    a. tree line

    b. forest line

c. frost line

d. timber line

2.4f   A soil augur is a device used to

a. collect a soil sample

b. measure the moisture content of a soil sample

c. weigh a soil sample

d. measure the organic content of a soil sample

2.5f   The permanently frozen soil found in the Arctic tundra is known as

a. moraine

b. glacial drift

c. permafrost

d. permasnow

2.6f   A Secchi disk is used to determine

a. the density of invertebrates in a soil sample

b. the turbidity or the extent of underwater visibility in a body of water, such as a lake

c. the intensity and duration of sunlight under a forest canopy

d. the abundance of flying insects

2.7f   Which of the following lists contains only the abiotic components of an ecosystem?

a. water, air, gravel, mosses, roads, buildings

b. highways, water, air, rocks, fungi, bridges

c. air, water, rocks, buildings, highways, soil

d. buildings, air, highways, pylons, gravel, water

2.8f   Which of the soil samples in Table 2.1 is/are acidic?

a. 1, 2 and 4

b. 2 and 4

c. 3 only

d. 3 and 5

Table 2.1

Sample number	pH
1	7.0
2	6.8
3	8.0
4	3.0
5	7.2

2.9f **A choice chamber is a piece of laboratory apparatus designed to study**

    a. plant growth under two different environmental conditions

    b. the food types selected by small mammals

    c. geotropisms in plants

    d. habitat preferences in small terrestrial invertebrates

2.10f **An anemometer may be used to measure**

    a. water flow in a river

    b. wind speed in a woodland

    c. soil depth in a field

    d. light levels under a forest canopy

2.11f **Which of the following could not be measured using a clinometer?**

    a. The height of a cliff

    b. The angle of a slope

    c. The girth of a tree

    d. The height of a tree

2.12f **A lotic environment consists of**

    a. standing water

    b. running water

    c. marine water

    d. polluted water

2.13f. The inside of a sampling tray used to examine inverte-brates collected from a pond or stream is most likely to be

a. black

b. white

c. brown

d. grey

2.14f The ratio of the quantity of water vapour in the air and the total amount of water vapour the same volume of air could hold if saturated, at the same temperature, is called the

a. maximum humidity

b. adjusted humidity

c. relative humidity

d. effective humidity

2.15f Which of the following might describe the aspect of a hillside?

a. A 35° gradient

b. Water-logged

c. North-facing

d. Tree-covered

2.16f The area between the high water mark and the low water mark on a beach is called the

a. benthic zone

b. intertidal zone

c. pelagic zone

d. strand line

2.17f In what type of physical environment would you expect to find a kelp forest?

a. High mountain

b. Saltwater

c. Freshwater

d. Semi-desert

2.18f A potometer is used to measure

    a. transpiration in plants

    b. respiration in small animals

    c. photosynthesis

    d. respiration in soil organisms

2.19f An atmometer is used to measure

    a. river flow

    b. atmospheric pressure

    c. soil moisture

    d. evaporation from a moist surface

2.20f The biological oxygen demand (BOD) of a sample of river water is a measure of the

    a. oxygen present

    b. organic pollution present

    c. nitrogen present

    d. chemical pollution present

Intermediate

2.1i Complete the following sentence: 'The loss in mass of dry soil that has been heated strongly would provide an estimate of its .... content.'

    a. water

    b. mineral

    c. inorganic

    d. humus

2.2i A dimictic lake is one in which

    a. the water is extremely shallow

    b. the water mixes from the bottom to the top during two periods of mixing each year

    c. the water mixes from the bottom to the top during a single period of mixing each year

    d. no thermal stratification occurs at any time of the year

2.3i **The thermocline in a thermally stratified lake is**

    a. a thin layer of water in which the temperature changes with depth more rapidly than in the layer above or the layer below

    b. located above the epilimnion

    c. located below the hypolimnion

    d. a thin layer of water in which the temperature changes with depth more slowly than in the layer above or the layer below

2.4i **When a temperature sensor is lowered into a thermally stratified lake, in which order will it pass through the various layers of water?**

    a. Hypolimnion, thermocline, epilimnion

    b. Epilimnion, thermocline, hypolimnion

    c. Thermocline, epiliminion, hypolimnion

    d. Epilimnion, hypolimnion, thermocline

2.5i **Frozen water has a density that is**

    a. the same as liquid water

    b. lower than liquid water

    c. slightly higher than liquid water

    d. much higher than liquid water

2.6i **Which of the following statements is true in relation to the tree line?**

    a. It may be defined in terms of altitude only

    b. It may be defined in terms of latitude only

    c. It may be defined in terms of altitude or latitude

    d. It is not affected by temperature

2.7i **The concept of pedogenesis relates to the mechanism of**

    a. ecological succession

    b. the decomposition of dead organisms

    c. soil formation

    d. soil loss due to erosion

2.8i   In a soil profile, the B horizon is the

a. topsoil

b. subsoil

c. organic matter

d. parent material

2.9i   Which of the following vertical zones in aquatic systems receives least light?

a. The euphotic zone

b. The eutrophic zone

c. The hypolimnion

d. The mesolimnion

2.10i   The infrared band in a false colour satellite image of part of northern Tanzania has been assigned to be red. This area contains a number of different habitat types. Plants strongly reflect infrared light so they appear dark red in this image. The dark red areas visible in the image are most likely to be

a. forest

b. forest or marshland

c. savannah or marshland

d. marshland

2.11i   A crown fire can only occur in a

a. forest

b. temperate grassland

c. savannah

d. peatland

2.12i   The capillary potential (water-retaining potential) of a soil is measured using a

a. potentiometer

b. tensiometer

c. hydrometer

d. hygrometer

2.13i **A vernier calliper could be used to measure**

    a. the length of a bird's wing

    b. rainfall

    c. the weight of a small mammal

    d. soil moisture

2.14i **The SI unit of illuminance is the**

    a. lux

    b. candela

    c. lumen

    d. foot candle

2.15i **Where would you locate an aquatic emergence trap?**

    a. At the bottom of a pond

    b. On the surface of the water in a pond

    c. 10cm below the surface of the water in a pond

    d. 10cm above the surface of the water in a pond

2.16i **Fig. 2.1 shows the structure of a net. What type of organisms would it be used to sample?**

    a. Flying insects

    b. Macrophytes

    c. Fishes

    d. Plankton

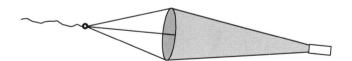

Fig. 2.1.

2.17i **A counting chamber marked with lines 3mm apart is most appropriate for counting**

    a. plankton

    b. pollen grains

c. bacteria

d. viruses

2.18i Changes in which of the following could not be monitored using an electronic data logger?

a. Air temperature

b. Biodiversity

c. The flow rate of water in a river

d. Precipitation

2.19i The flow of cold air down canyons and hillsides into valleys on calm, clear nights is called

a. cold-air drainage

b. cold-air conduction

c. a temperature inversion

d. a temperature cycle

2.20i The vertical structure of a soil may be divided into a series of horizons (vertical zones). From the top surface downwards, these are indicated by the abbreviations

a. O, A, B, C, R

b. R, C, B, A, O

c. A, B, C, O, R

d. O, R, A, B, C

Advanced

2.1a Which of the following is a possible effect of fire on terrestrial ecosystems?

a. The release of nutrients from the leaf litter to the soil

b. Herbivores gaining increased access to new green shoots

c. An increase in the abundance of fire-resistant species

d. All of the above

2.2a   **What sea temperature range can reef-building corals normally tolerate?**

a. 14–22°C

b. 17–26°C

c. 23–29°C

d. 25–40°C

2.3a   **Which of the following is used to record sound within a body of water (e.g. a lake)?**

a. A hygrophone

b. An aquafone

c. A microphone

d. A hydrophone

2.4a   **Which of the following is not part of the essential equipment required to undertake electrofishing (electric fishing) in a shallow river?**

a. An anode

b. A cathode

c. A sweep net

d. A boat

2.5a   **Which column in Table 2.2 accurately describes the labels in the storm hydrograph below (Fig. 2.2)?**

Table 2.2

Label	A	B	C	D
K	Time	Discharge	Discharge	Storm runoff
L	Discharge	Time	Time	Discharge
M	Base flow	Storm runoff	Base flow	Time
N	Storm runoff	Base flow	Storm runoff	Base flow

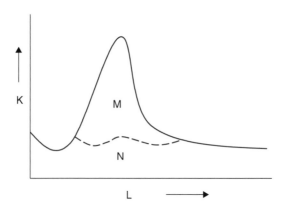

Fig. 2.2.

    a. A

    b. B

    c. C

    d. D

2.6a **Which of the following statements is false?**

    a. The specific heat of water is lower than that of land

    b. Water warms up more slowly than land

    c. The sea cools down more slowly than land

    d. Heat in water is distributed by horizontal and vertical mixing

2.7a **An isohyet is a line on a map than joins points of equal**

    a. temperature

    b. atmospheric pressure

    c. day length

    d. precipitation

2.8a **A line on a map that joins points of constant wind speed is called**

    a. an isotach

    b. an isobar

    c. an isomer

    d. an isograph

2.9a An isophene is a line on a map that joins neighbouring points of

    a. similar biological periodicity, such as the earliest date when individuals of a particular plant species flower

    b. similar vegetation height in a biome, such as the height of the canopy in a forest

    c. maximum high tide along a coastline

    d. similar biological diversity

2.10a The climate beneath the head of the vegetation is called the

    a. surface climate

    b. microclimate

    c. microthermal climate

    d. subterranean climate

2.11a The soil-forming process that is characteristic of the humid tropics in which organic materials do not accumulate at the soil surface, silica is leached downwards and iron oxides remain near the surface is called

    a. mineralisation

    b. melanisation

    c. podzolization

    d. lateralisation

2.12a Fig. 2.3 shows the pattern of rainfall (columns) and temperature (line) in which of the following biomes?

    a. Tundra

    b. Taiga

    c. Savannah

    d. Tropical rainforest

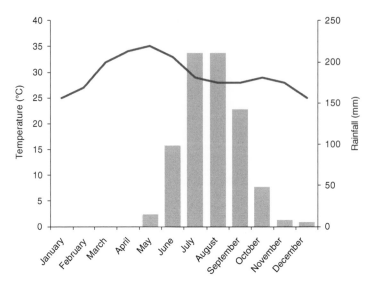

Fig. 2.3.

2.13a The albedo of a surface is the percentage of the incoming radiation that it reflects. Which section of Table 2.3 most accurately indicates the albedo values of the habitats listed?

Table 2.3

	A	B	C	D
Forest	18%	10%	10%	10%
Grass	10%	18%	18%	25%
Sand	25%	25%	80%	18%
Snow	80%	80%	25%	80%

a. A

b. B

c. C

d. D

2.14a The biological oxygen demand (BOD) of a sample of river water would be measured as follows:

a. Collect 2 samples of water. Sample 2 should be in a dark bottle. Measure dissolved O_2 (DO) in water sample 1; incubate sample

2 at 20°C for 5 days; measure DO in water sample 2. BOD (mg/L) = DO (mg/L) of first sample – DO (mg/L) of second bottle.

b. Collect 2 samples of water. Sample 2 should be in a dark bottle. Measure dissolved O_2 (DO) in water sample 1; incubate sample 2 at 20°C in the dark for 5 days; measure DO in water sample 2. BOD (mg/L) = DO (mg/L) of first sample – DO (mg/L) of second bottle.

c. Collect 2 samples of water. Sample 2 should be in a dark bottle. Measure dissolved O_2 (DO) in water sample 1; incubate sample 2 at 20°C in the dark for 10 days; measure DO in water sample 2. BOD (mg/L) = DO (mg/L) of first sample – DO (mg/L) of second bottle.

d. Collect 2 samples of water. Measure dissolved O_2 (DO) in water sample 1; incubate sample 2 at 25°C in the dark for 5 days; measure DO in water sample 2. BOD (mg/L) = DO (mg/L) of first sample – DO (mg/L) of second bottle.

2.15a **Which of the following laws describes the relationship between light intensity and the distance from the light source?**

a. Inverse law

b. Square law

c. Square-inverse law

d. Inverse-square law

2.16a **Acidity and alkalinity are measured on a pH scale: the negative logarithm (base 10) of the hydrogen ion concentration in solution. If sample A has a pH of 9.0 and sample B has a pH of 6.0, how much more acidic is sample B than sample A?**

a. 3 times

b. 30 times

c. 100 times

d. 1000 times

2.17a The constituents of soil samples can be sorted by particle size using a stack of nested sieves. Which sequence of mesh sizes (in microns) in Table 2.4 is correct?

Table 2.4

Sequence		A	B	C	D
Top sieve	1	4000	63	2000	4000
	2	2000	125	4000	2000
	3	250	250	500	500
	4	500	500	250	250
	5	63	126	125	125
Bottom sieve	6	125	63	63	63

a. A

b. B

c. C

d. D

2.18a A data buoy is most likely to be used to collect data

a. in the atmosphere

b. in the ocean

c. underground

d. in a forest canopy

2.19a Which of the following is the SI unit of radiation dose?

a. Sievert

b. Roentgen

c. Curie

d. Becquerel

2.20a A Burkard trap is used to monitor

a. heavy metals in river water

b. airborne insect numbers

c. airborne pollen and spores

d. plankton in the sea

3 Taxonomy and Biodiversity

This chapter contains questions about the system used to name organisms, their classification, and the diversity of life on Earth.

Foundation

3.1f **The binomial system of nomenclature was devised by Carl Linnaeus. He was a**

    a. Swedish physician

    b. Dutch botanist

    c. Swiss zoologist

    d. Austrian botanist

3.2f **A lichen is an association between**

    a. a green plant and a green alga

    b. a fungus and a protozoan or cyanobacterium

    c. a moss and a fungus

    d. a green alga or cyanobacterium and a fungus

3.3f **Which of the following statements about *Panthera pardus* and *Panthera leo* is false?**

    a. Both species are in the same class

    b. The two species are in different families

c. Both species are in the same genus

d. Both species are in the same order

3.4f **Which of the following classes contains the smallest number of extant (living) species?**

a. Reptilia

b. Amphibia

c. Aves

d. Mammalia

3.5f **Endemicity is the characteristic of being**

a. found only in one area

b. highly infectious

c. occurring over a wide geographical area

d. having high fecundity

3.6f **In the scientific name of an animal, the name indicating the species may begin with a capital (upper case) letter**

a. only if the name is derived from the name of a person (e.g. *Adamsi*)

b. only if the name is derived from that of a country (e.g. *Indicus*)

c. if the name is derived from that of a country or a person

d. under no circumstances

3.7f **The national bird of the United Kingdom is known as the European robin (*Erithacus rubecula*). This name is its**

a. binomial or scientific name

b. common or binomial name

c. scientific or vernacular name

d. vernacular or common name

3.8f **The suffix 'idae' in the name of a zoological taxon (e.g. Drosophilidae) indicates**

a. a genus

b. a class

    c. a family

    d. an order

3.9f **An identification key that allows an ecologist to identify a species (or other taxon) by asking a series of questions to which there are only two possible answers – for example, 'Does it have one pair of wings or two?'– is known as a**

    a. dichotomous key

    b. bifurcating key

    c. diverging key

    d. branching key

3.10f **Which of the following statements about the scientific name of an animal is false?**

    a. It is identical in all languages

    b. Once assigned it remains the same forever

    c. If it is a subspecies the name consists of three parts

    d. The name is based on Latin or Greek words

3.11f **Which section of Table 3.1 represents the correct sequence of taxonomic ranks used in zoology to classify animals (from the highest to the lowest)?**

Table 3.1

A	B	C	D
Phylum	Phylum	Phylum	Phylum
Class	Class	Order	Order
Order	Family	Family	Class
Family	Order	Class	Family
Genus	Genus	Genus	Genus
Species	Species	Species	Species

    a. A

    b. B

    c. C

    d. D

3.12f **The process by which the continents have moved apart over geological time, thereby determining the distribution of organisms, is known as**

    a. continental drift

    b. continental migration

    c. continental flow

    d. continental movement

3.13f **The application of distinctive names to each animal and plant is the definition of**

    a. nomenclature

    b. systematics

    c. classification

    d. taxonomy

3.14f **Herpetology is the scientific study of**

    a. amphibians

    b. reptiles

    c. amphibians and reptiles

    d. reptiles and birds

3.15f **Icthyology is the scientific study of**

    a. fishes

    b. insects

    c. birds

    d. fossils

3.16f **An expedition into an unexplored area of tropical forest consists of experts in entomology, mammalogy, herpetology and ornithology. Which of the following would this team of experts be least likely to be able to identify?**

    a. A rare species of frog

    b. A rare species of butterfly

    c. A rare species of rodent

    d. A rare species of freshwater crab

3.17f In which of the following areas do toads naturally occur?

a. New Zealand

b. Tasmania

c. Madagascar

d. Canada

3.18f A collection of preserved (dried) plants that is used for identification and scientific study is called

a. a herbarium

b. an arboretum

c. a flora

d. a conservatory

3.19f Ratites are

a. an extinct group of reptiles

b. a group of flightless birds

c. a type of coniferous tree

d. a type of protozoan

3.20f What type of coat would you expect the small African antelope *Cephalophus niger* to have based on its scientific name?

a. Striped

b. Brown

c. Spotted

d. Black

Intermediate

3.1i The animal phylum that contains the most extant species is the

a. Nematoda

b. Mollusca

c. Arthropoda

d. Annelida

3.2i **Which of the following statements about taxonomy is false?**

    a. In the species *Acinonyx jubatus soemmeringii* the last name indicates a subspecies

    b. In relation to any organism, the term 'scientific name' and 'vernacular name' mean the same thing

    c. The holotype of a species is the original specimen from which that species was first formally described

    d. The system of assigning two names to a species is known as the binomial system of nomenclature

3.3i **When a species comprises several subspecies the nominate subspecies is the one**

    a. with the largest geographical range

    b. that was described first

    c. that exhibits the greatest genetic variation

    d. that contains the largest number of individuals

3.4i **Which of the following statements about animal names is false?**

    a. The subspecies which was the first to be discovered has the species name repeated as its subspecific name, e.g. *Elephas maximus maximus*

    b. The scientific name of a species may be changed occasionally

    c. The vernacular name of a species may vary within and between countries

    d. Each genus of animals contains more than one species

3.5i **Which section of Table 3.2 ranks groups of animals correctly from the largest (most extant species) at the top, to the smallest (least extant species) at the bottom?**

Table 3.2

A	B	C	D
Insects	Mammals	Insects	Fishes
Fishes	Birds	Fishes	Insects
Birds	Fishes	Mammals	Birds
Mammals	Insects	Birds	Mammals

a. A

b. B

c. C

d. D

3.6i **When is it legitimate to abbreviate the name of the genus in the scientific name of an animal (e.g. to abbreviate *Loxodonta africana* to *L. africana*)?**

a. Never

b. When either the scientific name has previously been written out in full or when the name of the genus has previously been written out in full when referring to a different species of the same genus

c. Only when the scientific name has previously been written out in full

d. Only when the name of the genus has previously been written out in full when referring to a different species of the same genus

3.7i **How may new species be discovered?**

a. By assigning animals to new species when they had previously been classified as subspecies

b. By analysing the DNA of museum specimens that appear to be the same species on morphological evidence (i.e. they look the same)

c. By finding organisms in the wild that have not previously been described by scientists

d. All of the above

3.8i **Soil fauna may be classified by size. The organisms that are >2mm in size are collectively called the**

a. mesofauna

b. macrofauna

c. megafauna

d. microfauna

3.9i The abbreviation spp. after the name of the genus *Parus* (i.e. *Parus* spp.) means

a. a subspecies of *Parus*

b. an unspecified species of *Parus*

c. several unspecified species of *Parus*

d. several unspecified subspecies of *Parus*

3.10i The number of nematode species present in an ecosystem is difficult to determine because

a. they are all internal parasites of animals

b. they are all internal parasites of animals or plants

c. they are all microscopic in size

d. most species are small and live in the soil or inside other organisms

3.11i To which phylum does the species shown in Fig. 3.1 belong?

a. Porifera

b. Arthropoda

c. Cnidaria

d. Echinodermata

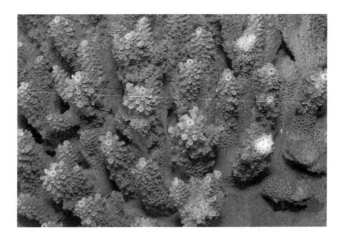

Fig. 3.1.

3.12i Which of the following countries has the largest number of flowering plant species?

a. Malaysia

b. Brazil

c. Ghana

d. Kenya

3.13i Which of the following taxa contains the largest number of living species?

a. Class Insecta

b. Phylum Arthropoda

c. Class Arachnida

d. Phylum Annelida

3.14i The documents that lay down rules for the naming of organisms are called

a. regulations

b. rules

c. codes

d. laws

3.15i A monotypic taxon is a taxonomic group that contains only one

a. genus

b. species

c. subspecies

d. immediately subordinate taxon

3.16i Dinosaurs appeared to vanish completely 65 million years ago. However, one lineage survived and evolved into the birds. The apparent extinction of a taxon is called

a. false extinction

b. pseudoextinction

c. fake extinction

d. near extinction

3.17i **Anemophilous plants are pollinated by**

    a. insects

    b. birds

    c. wind

    d. mites

3.18i **Which rule contends that fishes inhabiting cold waters tend to have more vertebrae than those living in warm waters?**

    a. Jordan's rule

    b. James's rule

    c. Jackson's rule

    d. Jewel's rule

3.19i **What is the word used for the combination of characteristics used to identify a bird species in the wild that consist of its size, relative proportions and its stance?**

    a. Jazz

    b. Jozz

    c. Jizz

    d. Jezz

3.20i **The study of undiscovered animals such as the yeti and Loch Ness monster is known as**

    a. pseudozoology

    b. cryptozoology

    c. cryozoology

    d. hyperzoology

Advanced

3.1a **A tropical forest community consists of a large number of species so the expected distribution of relative species abundance would be**

    a. normal

    b. negatively skewed

c. bimodal

d. log-normal

3.2a **The 'Wallace line' is a boundary between two regions of the world that contain distinctively different faunas and floras. It is located in**

a. Central America

b. West Africa

c. South America

d. Southeast Asia

3.3a **Who was the first person to employ rank-names to designate all natural groups of organisms of about the same status, for example, fishes?**

a. Carl Linnaeus

b. Charles Darwin

c. Aristotle

d. John Ray

3.4a **An ecologist discovered an unusual species of snail with a distinctive shell shape in a tropical forest in Ecuador. What type of specialist would you advise her to consult to determine whether or not it is a new species?**

a. An oncologist

b. A conchologist

c. An entomologist

d. A mycologist

3.5a **Table 3.3 shows the species present at two locations, site 1 and site 2 (indicated by √). Calculate the beta diversity index for these sites using the formula below:**

$$\text{Beta diversity index} = \frac{2c}{s1 + s2}$$

where,

$s1$ = number of species recorded at site 1

$s2$ = number of species recorded at site 2

c = number of species in common (i.e. recorded at both sites 1 and 2).

The value of the index is

a. 0.37

b. 0.46

c. 0.67

d. 0.72

Table 3.3

Species	Site 1	Site 2
A	✓	
B	✓	✓
C	✓	
D	✓	✓
E	✓	
F	✓	
G	✓	✓
H	✓	✓
I	✓	✓
J		✓

3.6a   The hippopotamus (*Hippopotamus amphibius*) belongs to the suborder Suiformes. In the binomial system of classification this falls between

a. the class Mammalia and the order Artiodactyla

b. the order Artiodactyla and the family Hippopotamidae

c. the family Hippopotamidae and the genus *Hippopotamus*

d. the suborder Tylopoda and the family Hippopotamidae

3.7a   Any one of two or more specimens of equal status upon which the original scientific description of a species is based is called a

a. holotype

b. lectotype

c. genotype

d. syntype

3.8a **The organisation responsible for producing the rules used in the naming of animals is the**

a. International Commission on Zoological Nomenclature

b. International Committee on Zoological Nomenclature

c. Interdisciplinary Commission on Zoological Names

d. International Agency for Zoological Nomenclature

3.9a **In zoological nomenclature, the second name in a trinomen indicates the**

a. genus

b. species

c. subspecies

d. subgenus

3.10a **In zoological nomenclature, the principle of priority essentially states that**

a. the valid name of a taxon is the oldest available name applied to it

b. the valid name of a species consists of two names

c. the name of each taxon must be unique

d. the first name of a species is always the same as the genus to which it belongs

3.11a **In zoological nomenclature, a taxon based on the fossilised work of an animal, for example, its fossilised trails, tracks or burrows, is called**

a. a palaeotaxon

b. a nominal taxon

c. a virtual taxon

d. an ichnotaxon

3.12a **A taxon with the suffix 'ini' (e.g. Bombini) denotes a**

a. tribe

b. subfamily

c. subgenus

d. race

3.13a The organisation responsible for overseeing matters of uniformity and stability in plant names is the

a. International Commission on Botanical Nomenclature

b. International Committee for Plant Classification

c. International Association for Plant Taxonomy

d. International Organisation for Plant Systematics

3.14a The latitudinal biodiversity gradient – the decline in the number of species from the Equator to the North and South Poles – was first recognised in the early 19th century by

a. Charles Darwin

b. Alfred Russel Wallace

c. Carl Bergmann

d. Alexander von Humboldt

3.15a The American entomologist Terry Erwin estimated the number of arthropod species in a tropical forest in Panama using the following information:

i. A total of 163 species of host-specific beetles are associated with a single tropical tree species (*Luehea seemannii*)

ii. Beetles represent approximately 40% of all arthropod species

iii. There are 50,000 species of tropical trees

iv. There are twice as many beetle species in the trees as on the ground

Approximately how many arthropod species are there in this tropical forest?

a. 20 million

b. 25 million

c. 30 million

d. 35 million

3.16a The data in Table 3.4 show the number of ant species recorded in various parts of the world. The pattern exhibited is known as

    a. a species diversity gradient

    b. a species abundance matrix

    c. a geographical gradient

    d. a species diversity matrix

Table 3.4

Country/US state	Number of ant species
Alaska	7
Brazil	222
Cuba	101
Iowa	73
Trinidad	134
Utah	63

3.17a The relationship between the area of an island and the number of species it contains may be studied by constructing a species-area curve similar to that in Fig. 3.2. On this graph the axes would be

    a. X = area; Y = species number

    b. X = area; Y = $\log_{10}$ species number

    c. X = $\log_{10}$ area; Y = $\log_{10}$ species number

    d. X = $\log_{10}$ area; Y = species number

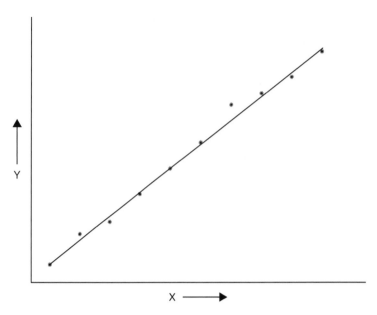

Fig. 3.2.

3.18a **Which classification rank occurs between kingdom and class in plants but not in animals?**

    a. Family

    b. Order

    c. Phylum

    d. Division

3.19a **Which of the following organisms are not classified as protists?**

    a. Amoebae

    b. Ciliates

    c. Diatoms

    d. Bacteria

3.20a **Which of the following groups of plants contains the largest number of living species?**

    a. Angiosperms

    b. Bryophytes

    c. Pteridophytes

    d. Gymnosperms

4 Energy Flow and Production Ecology

This chapter contains questions about the acquisition of energy by ecosystems, its fate within them, and the influence this has on ecosystem structure and functioning.

Foundation

4.1f The ultimate energy source in ecosystems is

    a. sometimes the sun and sometimes chemicals in the environment

    b. the heat produced by respiration

    c. always the sun

    d. always chemicals in the environment

4.2f The concept of a 'pyramid of numbers' was first proposed by

    a. Ernst Haeckel

    b. Eugene Odum

    c. Robert MacArthur

    d. Charles Elton

4.3f As energy moves along a food chain from one organism to another some of this energy is lost as the heat of

    a. respiration

    b. excretion

c. assimilation

d. digestion

4.4f   A predator at the top of a food chain may also be referred to as

a. an alpha predator

b. an apex predator

c. an apical predator

d. a climax predator

4.5f   An alternative name for a secondary producer is

a. a secondary consumer

b. a tertiary consumer

c. a principal producer

d. a primary consumer

4.6f   Which of the following elements is central to the structure of chlorophyll?

a. Selenium

b. Nickel

c. Magnesium

d. Copper

4.7f   The primary production of oak woodland excludes

a. the roots

b. the seeds

c. the bark

d. none of the above

4.8f   A piscivore feeds on

a. birds

b. small mammals

c. fishes

d. invertebrates

4.9f Which of the following statements is false?

a. All phototrophs are autotrophs

b. All heterotrophs are animals

c. All chemotrophs are autotrophs

d. Some organisms exhibit both autotrophic and heterotrophic modes of nutrition

4.10f The bacteria and fungi found in the soil are collectively known as

a. decomposers

b. detritivores

c. chemotrophs

d. primary producers

4.11f Consider the following food chain:

phytoplankton → zooplankton → fish → seal → killer whale

Which of the following statements about this food chain is false?

a. As energy moves along this food chain some energy is lost as the heat of respiration between the phytoplankton and the killer whale at three points only

b. The fish is a secondary consumer

c. The ultimate source of the energy for this food chain is sunlight

d. There must be more zooplankton than seals and fewer killer whales than fish in this food chain

4.12f Which of the following statements is false?

a. Net primary productivity may be measured as units of mass/unit area/unit time interval

b. Chemotrophs use chemicals to trap energy from the sun

c. Secondary production is the production of chemical energy by animals

d. Primary production is the production of chemical energy from light

4.13f **What types of organisms are included in the trophic level designated T$_4$?**

a. Green plants

b. Top carnivores

c. Bacteria

d. Decomposers

4.14f **An abstract representation of the major feeding relationships among the producers and consumers in an ecosystem is best described as a**

a. feeding network

b. feeding lattice

c. food web

d. food chain

4.15f **Which of the following terms most accurately describes the manner in which energy moves within ecosystems?**

a. Flows

b. Cascades

c. Oscillates

d. Circulates

4.16f **A sanguivore feeds on**

a. plant sap

b. blood

c. faeces

d. fungi

4.17f **Ram feeding is practised by**

a. big cats while eating a large quantity of meat over a short period

b. large predatory fishes while swimming forward with their mouths open wide

c. elephants while consuming large quantities of browse

d. humming birds while they are feeding on nectar from flowers

4.18f The energy entering and leaving the body of a mammal is shown by the arrows in Fig. 4.1. The relative thickness of each arrow is intended to represent the quantity of energy involved. Which of the following statements about this diagram is true?

    a. Respiratory losses are over-represented in the diagram

    b. Urine does not have an energy content

    c. Two important energy losses have been omitted from the diagram

    d. The thicknesses of the arrows do not accurately reflect the relative sizes of the energy flows

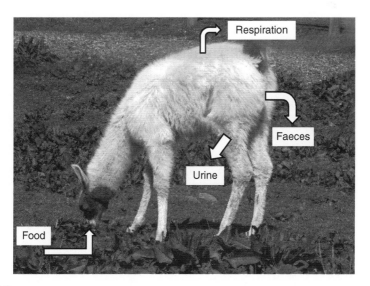

Fig. 4.1.

4.19f The photograph below (Fig. 4.2) is of the skull of a mammal. To which trophic level does this species belong?

    a. T_1

    b. T_2

    c. T_3

    d. T_4

Fig. 4.2.

4.20f Predators that feed on only a few prey types are called

    a. oligophagous

    b. monophagous

    c. polyphagous

    d. oligotrophic

Intermediate

4.1i The primary productivity of a forest ecosystem is best measured by calculating the

    a. wet biomass of animals/unit area/unit time

    b. wet biomass of green plants/unit area/unit time

    c. dry biomass of animals/ unit area/unit time

    d. dry biomass of green plants/unit area/unit time

4.2i Which of the following pairs of terms do not mean the same thing?

    a. Trophic level 1 and primary producer

    b. Primary consumer and secondary producer

    c. Tertiary producer and secondary consumer

    d. Detritivore and decomposer

4.3i **Energy flow in a generalised food chain is shown in Fig. 4.3. Which of the following statements about this diagram is true?**

a. G represents energy loss via respiration and F represents decomposers

b. B represents green plants and D represents top carnivores

c. A represents energy from the sun and E represents decomposers

d. C represents herbivores and G represents decomposers

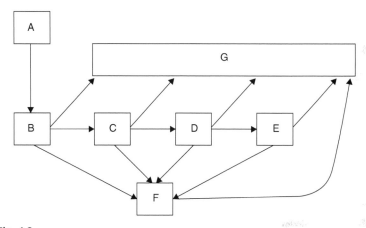

Fig. 4.3.

4.4i **Which of the following statements is true for the pyramid of numbers in Fig.4.4 ?**

a. A = decomposers and B = primary producers

b. A = herbivores and D = carnivores

c. B = energy from the sun and C = herbivores

d. C = decomposers and E = top carnivores

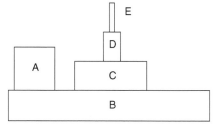

Fig. 4.4.

4.5i   **Which of the following statements about the pyramid of numbers shown in Fig 4.5 is false?**

a. The primary producers are less common than the carnivores

b. A could represent trees, B could represent herbivorous insects and C could represent bacteria living inside the insects

c. A could represent grasses, B could represent gazelles and C could represent lions

d. The pyramid could represent a parasitic food chain

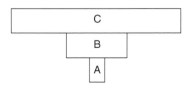

Fig. 4.5.

4.6i   **Which of the following statements is true?**

a. Gross secondary production – respiration = net primary production

b. Gross secondary production + respiration = net secondary production

c. Gross primary production – respiration = net primary production

d. Net primary production – respiration = gross primary production

4.7i   **The expression below is used to calculate**

a. digestive efficiency

b. feeding efficiency

c. conversion efficiency

d. consumption efficiency

$$\frac{intake\ at\ trophic\ level\ n}{net\ productivity\ at\ trophic\ level\ n-1}$$

4.8i   **Which of the following statements is false?**

a. All pyramids of numbers are wider at the bottom than at the top

b. Parasitic food chains have a pyramid of numbers that is inverted

    c. Pyramids of energy are always narrower at the top than at the bottom

    d. All pyramids of biomass are narrower at the top than at the bottom

4.9i **Which biome has the highest net primary production per unit area?**

    a. Desert scrub

    b. Temperate grassland

    c. Tropical moist green forest

    d. Temperate deciduous forest

4.10i **Which of the following food chains would exhibit an inverted pyramid of numbers?**

    a. grasses → gazelles → cheetahs

    b. phytoplankton → zooplankton → fishes → penguins

    c. trees → aphids → insect-eating birds

    d. host → parasites → hyperparasites

4.11i **What does the letter X represent in the expression below?**

    a. H_{10}

    b. H_{12}

    c. H_6

    d. H_8

$$6CO_2 + 6H_2O \rightarrow C_6XO_6 + 6O_2$$

4.12i **The partitioning of, or balance sheet for, the energy obtained by an animal (a bullock), and the quantities of energy used for various biological functions are shown in Table 4.1. This is called**

    a. an activity budget

    b. an energy flow diagram

    c. an energy budget

    d. an energy spreadsheet

Table 4.1

	kJ/m²/yr
Food eaten	3052
Assimilated	1145
Heat lost in respiration	1021
Egesta	1908
Secondary production	126

4.13i **Comparing the primary productivity of terrestrial ecosystems by measuring the dry biomass of the aerial parts of the plants is misleading because**

   a. the roots are also part of the production and some plant species produce more roots relative to aerial parts than others

   b. the roots are also part of the production and all plant species produce the same amount of root material relative to aerial parts

   c. measuring the dry biomass does not take into account the mass of water in the plants

   d. it does not take into account the effects of environmental conditions

4.14i **The net energy gain from eating a particular food source may be expressed as**

$$\frac{energy\ intake\ of\ the\ forager}{time\ invested\ in\ foraging}$$

This is known as its

   a. efficiency

   b. profitability

   c. assimilation efficiency

   d. gainfulness

4.15i **Calculate the gross assimilation efficiency of a herbivore using the following data (Table 4.2) collected over a period of one week.**

Table 4.2

	Wet weight (kg)	Water (%)
Food consumed	119	70
Dung produced	139	81

a. 21%

b. 26%

c. 29%

d. 34%

4.16i The equation below shows a type of

a. photosynthesis

b. respiration

c. assimilation

d. chemotrophism

$$4FeCO_3 + O_2 + 6H_2O \rightarrow 4Fe(OH)_3 + 4CO_2 + Energy$$

4.17i Charles Elton noted that the number of links in a food chain rarely exceeds

a. 3

b. 5

c. 7

d. 9

4.18i Which of the following factors affects the digestibility of grass?

a. The species of grass

b. The time in its growing season when the grass is eaten

c. The species of herbivore that has consumed it

d. All of the above

4.19i Animals that engage in coprophagy eat

a. faeces

b. insects

c. detritus

d. plankton

4.20i   **Most mammals have a limited ability to digest plant material because of their inability to produce**

   a. fructase

   b. cellulase

   c. lipase

   d. sucrase

Advanced

4.1a   **Entropy is a measure of the lack of order or randomness in a system. In nature there is a gradual increase in randomness (high disorder) as complex organic molecules are broken down into their constituent parts when organisms die and decay. The process of photosynthesis in the cells of green plants causes**

   a. no change in entropy

   b. an increase in entropy followed by a decrease in entropy

   c. an increase in entropy

   d. a decrease in entropy

4.2a   **In ecosystems, complexity of food webs tends to lead to**

   a. instability

   b. stability

   c. extinction

   d. speciation

4.3a   **Any route by which chemical energy contained within detrital organic carbon becomes available to the biota may be called**

   a. a detritus food chain

   b. a grazing food chain

   c. chemosynthesis

   d. a nutrient cycle

4.4a What proportion of the total energy in incoming solar radiation ends up in the net primary production in forests?

    a. 5.0–10.0%

    b. 3.0–5.0%

    c. 1.0–3.0%

    d. 0.5–1.0%

4.5a Voles are omnivorous small mammals. If 60% of the mass of a vole's diet is plant material and the remainder consists of primary consumers, how should the energy in a vole population be assigned to trophic levels when constructing a pyramid of energy?

    a. 40% primary producer and 60% secondary consumer

    b. 60% primary consumer and 40% secondary consumer

    c. 100% primary consumer

    d. 40% secondary consumer and 60% tertiary consumer

4.6a An organism described as xylophagous feeds on

    a. leaves

    b. dead animals

    c. fruit

    d. wood

4.7a The expression below is used in production ecology and is known as

    a. gross assimilation efficiency

    b. growth efficiency

    c. Lindeman's efficiency

    d. consumption efficiency

$$\frac{assimilation\ at\ trophic\ level\ n}{assimilation\ at\ trophic\ level\ n - 1}$$

4.8a The fate of the energy (units = x10^3 Joules) taken in by a small mammal in a day is shown in Fig. 4.6. What percentage of the energy in the food consumed is lost?

    a. 85.8%

    b. 12.1%

c. 97.9%

d. 93.1%

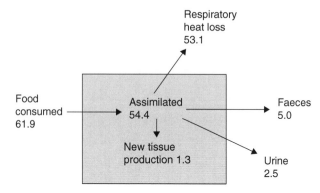

Fig. 4.6.

4.9a **What is the relationship between annual gross primary productivity and evapotranspiration?**

a. Strong positive correlation

b. Weak positive correlation

c. Negative correlation

d. No correlation

4.10a **The fate of energy (kJ/m²/yr) falling on an old field eco-system in Michigan in the United States is shown in Fig. 4.7. Approximately what percentage of the total insolation is used to produce new plant material?**

a. 0.1%

b. 1.0%

c. 2.0%

d. 5.0%

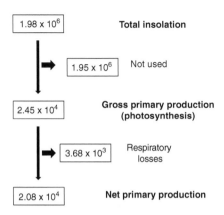

Fig. 4.7.

4.11a **Which of the following radioactive isotopes have been used to study trophic structures and energy pathways in ecosystems?**

    a. ^{13}C and ^{15}N

    b. ^{13}C and ^{34}S

    c. ^{34}S and ^{15}N

    d. ^{13}C, ^{15}N and ^{34}S

4.12a **Energy absorbed from the sun by plants is converted into glucose and this is used in cellular respiration to produce**

    a. ADP

    b. ATP

    c. DDT

    d. DDE

4.13a **Unit leaf rate (ULR) is the rate of increase in dry weight of a plant per unit of leaf area ($g/m^2/day$).**

ULR x leaf area index =

    a. ecological growth efficiency

    b. photosynthetic efficiency

    c. gross primary productivity

    d. net primary productivity

4.14a Calculate the volume of wood (V) in the trunk of a tree of height (h) = 23m, whose radius at breast height (r) = 0.5m, using the formula below:

$$V(m^3) = \frac{\pi r^2 h}{2}$$

a. 8.2 m³

b. 9.0 m³

c. 10.1 m³

d. 11.5 m³

4.15a Which of the following methods has not been used to trace food chains?

a. Radioactive labelling with phosphorus-32

b. The precipitin test (antigen-antibody reaction)

c. Gut content analysis

d. They have all been used

4.16a Meat production data from cattle and rabbits are compared in Table 4.3.

Table 4.3

	1 cow	300 rabbits
Total body weight	590kg	590kg
Food consumed/day	7.56kg	30.24kg
Duration of I ton of hay	120 days	30 days
Heat loss/day	83,680 kJ	334,720 kJ
Gain in weight/day	0.91kg	3.63kg
Gain from 1 ton of food	108.86kg	108.86kg

Which of the following statements is true?

a. Meat production from the rabbits is less efficient because they lose more energy in the form of heat than the cow

b. The efficiency of meat production is the same in both systems

c. Meat production in the cow is more efficient because it eats less food per day than the rabbits

d. Meat production is more efficient in the rabbits because they produce it four times as quickly as the cow from the same quantity of hay

4.17a If the time taken for food to pass through the gut of an animal is 8 hours and the dry weight of the gut contents is 7.3kg, what is the dry weight of food consumed per day?

a. 21.9kg

b. 14.6kg

c. 16.5kg

d. 29.2kg

4.18a The data in Table 4.4 show oxygen consumption by two species of spiders during the day and night.

Table 4.4

Species	Oxygen consumption (mm^3/mg/h)	
	Day	Night
M	0.63	0.79
O	0.94	1.34

Which of the following statements is supported by the data?

a. Both species are twice as active at night than during the day

b. Species O consumes less food than Species M

c. Species M is slightly more active at night while species O is much more nocturnal

d. Species O is equally active during the day and night while species M is more nocturnal

4.19a The most photosynthetically efficient types of plants are those that are referred to as

a. C2

b. C3

   c. C4

   d. C5

4.20a What is X in the equation below?

   a. Net growth efficiency

   b. Gross growth efficiency

   c. Food chain efficiency

   d. Gross ecological efficiency

$$X = \frac{energy\ used\ in\ growth}{energy\ assimilated}$$

5 Nutrient and Material Cycles

This chapter contains questions about the role of nutrients in organisms, their cycling between the biological and physical components of the environment, and the hydrological cycle.

Foundation

5.1f **The organism shown in Fig. 5.1 is a**

    a. saprotroph

    b. microphyte

    c. mesophyte

    d. phytotroph

5.2f **Excessive algal growth occurs when large quantities of nutrients are washed into freshwater ecosystems, especially ponds and lakes. This process is called**

    a. ammonification

    b. eutrophication

    c. denitrification

    d. acidification

Fig. 5.1.

5.3f **The most commonly found elements in general plant fertilisers are**

    a. iron, nitrogen and zinc

    b. phosphorus, zinc and magnesium

    c. nitrogen, potassium and phosphorus

    d. zinc, manganese and magnesium

5.4f **During heavy rainfall nutrients may be washed out of the top layers of the soil in a process known as**

    a. nutrient washing

    b. leaking

    c. leaching

    d. denitrification

5.5f **In some countries guano is an important source of phosphate fertiliser that is produced naturally by**

a. insects and other invertebrates

b. seabirds and bats

c. marine fishes and crustaceans

d. terrestrial mammals

5.6f **Which of the following elements is most important in egg shell production in birds?**

a. Magnesium

b. Potassium

c. Calcium

d. Phosphorus

5.7f **Nitrogen-fixing bacteria occur in the root nodules of**

a. grasses

b. cacti

c. conifers

d. legumes

5.8f **Most plant nutrients are optimally available at pH 6.5 – 7.5. Nutrients are absorbed through the roots in the soil water. Liming an acidic soil may increase crop yield because**

a. it increases the soil pH, making nutrients more available

b. it decreases the soil pH, making nutrients more available

c. it helps the soil retain water, making it easier for plants to absorb nutrients

d. the calcium carbonate in the lime interacts with the soil nutrients, increasing the rate of uptake

5.9f **Which of the following pathways was most recently introduced into the carbon cycle?**

a. C in plant sugars → C in animal sugars

b. C in atmospheric CO_2 → uptake by plants and use in photosynthesis

c. Combustion of fossil fuels → release of CO_2 to atmosphere

d. C in dead organic matter in sea → marine sediments

5.10f Which section of Table 5.1 accurately represents the chemical composition of the atmosphere?

Table 5.1

	A	B	C	D
Oxygen	21%	21%	78%	78%
Carbon Dioxide	0.93%	0.04%	0.04%	0.93%
Nitrogen	78%	78%	21%	21%
Argon	0.04%	0.93%	0.93%	0.04%

a. A

b. B

c. C

d. D

5.11f A nutrient required by a plant in very small quantities is called a

a. miniscule element

b. minor element

c. minute element

d. trace element

5.12f Nutrients enter ecosystems from

a. the atmosphere

b. the breakdown of rocks and soil

c. the decomposition of dead organisms

d. all of the above

5.13f Which of the following processes is not part of the water cycle?

a. precipitation

b. ammonification

c. evapotranspiration

d. infiltration

5.14f **During a storm in a coniferous forest much of the precipitation reaches the ground as 'indirect throughfall' consisting of**

    a. leaf drips

    b. stem flow

    c. run off

    d. leaf drips and stem flow

5.15f **The upper surface of the groundwater is called the**

    a. groundwater level

    b. field capacity

    c. water table

    d. water capacity

5.16f **Phosphorus is extracted from rock and used to fertilise crops. Much of the phosphorus that is not taken up by plants eventually ends up in**

    a. the atmosphere

    b. the sea

    c. the soil

    d. the secondary producers

5.17f **'Mull' and 'mor' are types of**

    a. humus

    b. soil

    c. litter

    d. detritus

5.18f **The absorption of phosphate, ammonium, nitrate and potassium ions by some plants is aided by mycorrhizae. A mycorrhiza is an association between living plant root cells and**

    a. a bacterium

    b. a fungus

    c. an alga

    d. a liverwort

5.19f **Which of the following elements is least abundant in living things?**

    a. Nitrogen

    b. Oxygen

    c. Hydrogen

    d. Sulphur

5.20f **Which of the following is the most abundant ion dissolved in seawater?**

    a. Sodium

    b. Calcium

    c. Chlorine

    d. Magnesium

Intermediate

5.1i **Transpiration, evapotranspiration and evaporation are processes that occur in the water cycle. Which of the following relationships between these processes is true?**

    a. Transpiration − evapotranspiration = evaporation

    b. Evaporation + transpiration = evapotranspiration

    c. Evapotranspiration + evaporation = transpiration

    d. Evaporation − evapotranspiration = transpiration

5.2i **Hydrothermal vents are important sources of nutrients for food chains in**

    a. rivers and volcanic lakes

    b. deep oceans only

    c. estuaries and deep oceans

    d. volcanic lakes and deep oceans

5.3i **Tiny particles, especially dead organic matter, that sink from the upper to the deep parts of the ocean are collectively known as**

    a. marine detritus

    b. marine snow

    c. marine litter

    d. marine drift

5.4i   **In what form is nitrogen absorbed from the soil by plants?**

    a. nitrate, nitrite and ammonium

    b. nitrate and ammonium

    c. ammonium only

    d. nitrate only

5.5i   **All fungi absorb sugars and other organic molecules through their cell walls and cell membranes. They obtain their food by feeding as**

    a. saprobes

    b. saprobes and symbionts

    c. saprobes and parasites

    d. saprobes, symbionts and parasites

5.6i   **Where would you expect to find catalytic soil enzymes?**

    a. On the surface membranes of viable cells

    b. Complexed in the soil matrix

    c. In the soil solution

    d. All of the above

5.7i   **In waterlogged soil, denitrifying bacteria**

    a. breakdown nitrates and release nitrogen to the air

    b. absorb nitrites from the soil

    c. convert ammonia to nitrates

    d. absorb nitrogen from the air

5.8i   **The process shown in the expression below occurs in the atmosphere and is caused by**

    a. lightning

    b. sunlight

c. heat

d. chemical pollution

Nitrogen + oxygen → Nitrogen oxides

5.9i **Approximately how much of the organic carbon in a tropical rainforest would you expect to find held in its biomass?**

a. 35%

b. 40%

c. 50%

d. 80%

5.10i **In some parts of the ocean the action of the wind causes nutrient-rich water to move from depth to the surface. This movement is known as**

a. upsurging

b. an up-current

c. upwelling

d. an updraft

5.11i **Which of the following is a micronutrient for plants?**

a. Potassium

b. Phosphorus

c. Nitrogen

d. Zinc

5.12i **The main reservoir of carbon available to plants is**

a. glucose

b. carbon dioxide

c. calcium carbonate

d. cellulose

5.13i **Which of the following sequences accurately represents the movement of a single carbon atom in the carbon cycle?**

    a. carbon dioxide in the air → plant carbon → photosynthesis → animal carbon → respiration → carbon dioxide in the air

    b. carbon dioxide in the air → photosynthesis → plant carbon → animal carbon → respiration → carbon dioxide in the air

    c. carbon dioxide in the air → photosynthesis → plant carbon → respiration → animal carbon → carbon dioxide in the air

    d. carbon dioxide in the air → photosynthesis → plant carbon → animal carbon → carbon dioxide in the air → respiration

5.14i **On the continents, the largest quantity of water is stored in the**

    a. subsurface

    b. glaciers

    c. lakes

    d. soil and subsoil

5.15i **Most of the carbon on Earth is stored in the**

    a. hydrosphere

    b. lithosphere

    c. atmosphere

    d. soil

5.16i **Which of the following statements about the nitrogen cycle is false?**

    a. Ammonia is converted to nitrite by nitrification

    b. Organic nitrogen in dead organisms is converted to ammonium by assimilation

    c. Atmospheric nitrogen is fixed by bacteria in the root nodules of some plants

    d. Nitrates are converted to atmospheric nitrogen by denitrification

5.17i Water evaporates over the oceans and precipitates over land as a result of

a. deposition

b. advection

c. sublimation

b. infiltration

5.18i Water is held in various components of the water cycle known as 'reservoirs'. The time it spends in a reservoir is called the residence time. Which of the following reservoirs has the longest residence time?

a. The ocean

b. The atmosphere

c. The groundwater

d. The glaciers

5.19i The important components and processes involved in the nitrogen cycle are shown in Fig. 5.2. Which organisms are represented by the letters X and Y?

a. X = Nitrobacter, Y = Nitrosomonas

b. X = Nitrosomonas, Y = Nitrobacter

c. X = Rhizobium, Y = Nitrobacter

d. X = Nitrosomonas, Y = Rhizobium

5.20i That part of the total precipitation that is useful to plants is called the

a. effective precipitation

b. affective precipitation

c. obtainable precipitation

d. available precipitation

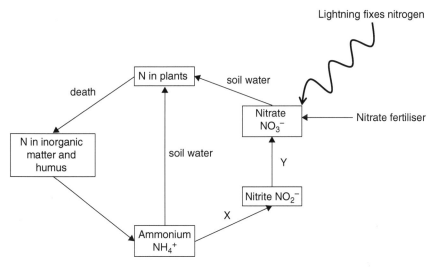

Fig. 5.2.

Advanced

5.1a When a stream's energy resources are derived from outside the stream itself they are referred to as

    a. autotrophic

    b. autochthonous

    c. allochthonous

    d. allopatric

5.2a What is the name of the process illustrated below?

$$\text{Ammonium } NH_4^+ \longrightarrow \text{Nitrite } NO_2^-$$

    a. Nitrification

    b. Denitrification

    c. Ammonification

    d. Nitrogen fixation

5.3a   **In leaf litter, the term 'mast' refers to**

    a. the remains of dead insects

    b. decomposing pine cones

    c. pieces of decomposing tree bark

    d. acorns and other seeds, and nuts

5.4a   **In a large lake, which of the following is likely to be the most important nitrogen sink?**

    a. The large fish

    b. The outflow

    c. The phytoplankton

    d. Sedimentation

5.5a   **In which of the following climatic zones would you expect the longest half-life of nitrogen in leaf litter to be found?**

    a. Boreal

    b. Polar

    c. Temperate

    d. Subtropical

5.6a   **Biogeochemical cycles may be described in terms of pools (the quantity of a particular chemical substance in a specific ecosystem component) and flux rates (the quantity of material moving between them). Which of the following would be the most appropriate general measure of flux rate?**

    a. Units/area/day

    b. Units/day

    c. Units/volume

    d. Units/area or volume/day

5.7a   **The Hubbard Brook study examined the effect of deforestation on the chemistry of stream water by comparing streams in deforested (experimental) and undisturbed**

(control) watersheds. In the experimental watershed, woody vegetation was cut and left in place, and the growth of higher plants was inhibited by herbicide. After deforestation, which of the following nutrients increased in concentration in the stream in the experimental watershed?

a. NO_3^- and Ca^{++}

b. Ca^{++} and Mg^{++}

c. Mg^{++}, NO_3^- and Ca^{++}

d. NO_3^-, Ca^{++}, K^+ and Mg^{++}

5.8a The biogeochemical relationships for a nutrient in a simple aquatic ecosystem (a pond of area 4 acres) have been described by Collier *et al.* (1974) and are illustrated in Fig. 5.3. The boxes represent nutrient pools and the arrows represent fluxes between the pools.

$$\text{The turnover rate} = \frac{the\ flux\ rate\ of\ the\ nutrient\ into\ or\ out\ of\ a\ pool}{the\ quantity\ of\ nutrient\ in\ the\ pool}$$

Which of the following has the highest turnover rate?

a. The flux from the water mass to the producers

b. The flux from the producers to the sediments

c. The flux from the producers to the heterotrophs

d. The flux from the sediments to the water mass

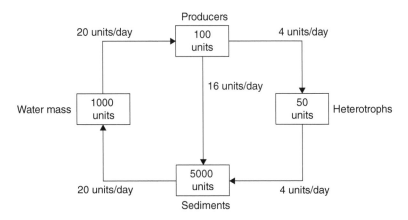

Fig. 5.3.

5.9a Sewage sludge cake is used as a soil conditioner to increase the nitrogen content of agricultural soils. Table 5.2 shows the quantity of nitrogen in the soil after annual applications of sludge for 10 years. The total available in any year can be determined by adding up all the values in each column (the amount added that year and the amounts remaining from applications in previous years).

Table 5.2

Year of Application	Year after application									
	1	2	3	4	5	6	7	8	9	10
1	38	19	10	5	2	1	0	0	0	0
2	-	38	19	10	5	2	1	0	0	0
3	-	-	38	19	10	5	2	1	0	0
4	-	-	-	38	19	10	5	2	1	0
5	-	-	-	-	38	19	10	5	2	1
6	-	-	-	-	-	38	19	10	5	2
7	-	-	-	-	-	-	38	19	10	5
8	-	-	-	-	-	-	-	38	19	10
9	-	-	-	-	-	-	-	-	38	19
10	-	-	-	-	-	-	-	-	-	38
Total N kg/ha										

What happens to the quantity of available nitrogen in the soil in the years following the sixth application?

a. It continues to rise with each subsequent application

b. It begins to decrease

c. It stabilises

d. It becomes unpredictable

5.10a The Sargasso Sea is an area of very low productivity in the Atlantic Ocean where the surface waters are very low in nutrients. Primary productivity can be measured from data on the uptake of carbon dioxide. A series of 3-day enrichment experiments conducted by Menzel and Ryther (1961) produced the results shown in Table 5.3.

Table 5.3

Culture	Nutrients added	Relative uptake of ^{14}C
Control	None	100%
Experimental 1	N + P + metals	1290%
Experimental 2	N + P	110%
Experimental 3	N + P + metals except Fe	108%
Experimental 4	N + P + Fe	1200%

What factor appears to be most important in limiting primary production?

a. Iron

b. Nitrogen

c. Phosphorus

d. A metal other than Iron

5.11a Which of the following is a major controlling factor for the growth of diatom populations in freshwater ecosystems?

a. Sulphur

b. Silicon

c. Carbon

d. Cobalt

5.12a What percentage of the richest ocean water is nitrogen?

a. 0.00005%

b. 0.005%

c. 0.05%

d. 0.5%

5.13a **Many studies have found no relationship between soil nu-trients and forest growth. The most likely explanation for this is that**

    a. there were errors in the soil sampling methodology

    b. the method of soil analysis was not sensitive enough to distin-guish soils with high concentrations of nutrients from those with low concentrations

    c. the method used to measure forest growth was flawed

    d. soil nutrients may be present but unavailable to plants

5.14a **Which of the following statements is true for a temperate lake in summer?**

    a. The metalimnion is anoxic and the hypolimnion is nutrient poor

    b. The hypolimnion is nutrient poor and epilimnion is anoxic

    c. The epilimnion is nutrient poor and the hypolimnion is anoxic

    d. The thermocline is anoxic and the metalimnion is nutrient poor

5.15a **Nitrogenase enzymes are responsible for the process of**

    a. nitrification

    b. nitrogen fixation

    c. denitrification

    d. ammonification

5.16a **Very small pellets of faecal material produced by small soil invertebrates are called**

    a. crass

    b. wrasse

    c. frass

    d. trass

5.17a **Which of the following expressions is false?**

    a. Evapotranspiration = precipitation – runoff – percolation

    b. Evaporation from soil + plant transpiration = evapotranspiration

   c. Evapotranspiration = precipitation + runoff – percolation

   d. Evapotranspiration + runoff = precipitation – percolation

5.18a The Haber-Bosch process is used to produce fertiliser from

   a. the sulphur in volcanic rocks

   b. the nitrogen in the atmosphere

   c. the phosphorus in animal waste

   d. the potassium in sedimentary rocks

5.19a The effect of the quantity of an essential nutrient in the environment on the growth of an organism is illustrated in Fig. 5.4. The regions A, B and C represent

   a. A = nutrient sufficient; B = nutrient limiting; C = nutrient toxic

   b. A = nutrient toxic; B = nutrient limiting; C = nutrient sufficient

   c. A = nutrient limiting; B = nutrient toxic; C = nutrient sufficient

   d. A = nutrient limiting; B = nutrient sufficient; C = nutrient toxic

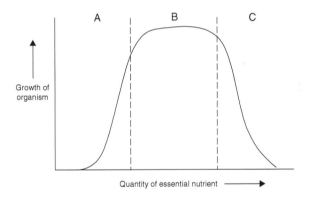

Fig. 5.4.

5.20a Which of the following is not associated with good aeration of soil?

   a. Good drainage

   b. Slow oxidation of humus

   c. Good plant root penetration

   d. A good crumb structure

6 Ecophysiology

This chapter contains questions about the physiological adaptations of animals and plants to their environment

Foundation

6.1f Homeothermic animals are able to regulate their internal body temperature physiologically and have therefore been able to colonise a wide range of habitats. They are all

a. mammals

b. mammals or birds

c. mammals, birds or reptiles

d. birds or reptiles

6.2f A deciduous tree

a. loses its leaves in winter

b. can only survive in temperate latitudes

c. always has needle-shaped leaves

d. always produces seeds within cones

6.3f An ecologist travelling from the Arctic Circle to the Equator noticed that the ears of the various local fox species tended to increase in size in relation to the size of the head as she progressed southward. This phenomenon is described in

   a. Allee's rule

   b. Allen's rule

   c. Alwin's rule

   d. Albert's rule

6.4f A xerocole is an animal adapted to living

   a. in saltwater

   b. underground

   c. in low temperatures

   d. in deserts

6.5f The maintenance within an organism or group of organisms of a steady state by means of physiological and behavioural feedback loops is known as

   a. homeostasis

   b. homeothermy

   c. homology

   d. homoplasy

6.6f Which of the following terms could not be applied to a large turtle?

   a. Gigantotherm

   b. Endothermic homeotherm

   c. Inertial homeotherm

   d. Ectothermic homeotherm

6.7f Plants described as succulents are most likely to occur in

   a. tropical forests

   b. freshwater swamps

c. temperate grasslands

d. hot deserts

6.8f **A plant species that has a narrow range of tolerance to an environmental variable, such as temperature, salinity or acidity, is referred to as being**

a. stenoecious

b. dioecious

c. monoecious

d. euryecious

6.9f **Gular fluttering is a rapid movement of the throat which increases evaporation from expired air and thereby heat loss in**

a. desert mammals

b. birds

c. reptiles

d. amphibians

6.10f **Osmoconformers are animals whose body fluid concentration is exactly the same as their immediate environment, for example marine invertebrates. Those osmoconformers that are able to tolerate wide changes in the osmotic concentration of their immediate environment are referred to as being**

a. euryhaline

b. stenohaline

c. isohaline

d. hyperhaline

6.11f **In the nephrons of desert rodents the Loop of Henlé is exceptionally**

a. short

b. thick

c. straight

d. long

6.12f **Some insects (e.g. cockroaches) deposit uric acid around the body, especially in the cuticle, to**

    a. save water

    b. save energy

    c. conserve body heat

    d. strengthen their exoskeleton

6.13f **An animal that maintains the osmotic concentration of its body fluids at a higher concentration than that of its immediate environment is described as a**

    a. hypoosmotic conformer

    b. hypoosmotic regulator

    c. hyperosmotic regulator

    d. hyperosmotic conformer

6.14f **The ability of an animal to alter the range over which a particular physiological variable is maintained is called**

    a. acclimatisation

    b. homeostasis

    c. adjustment

    d. accommodation

6.15f **Some amphibians have evolved vascularised hairs on the skin whose purpose is to**

    a. increase the surface area available for gaseous exchange

    b. assist in temperature regulation

    c. facilitate efficient removal of excretory products

    d. attract members of the opposite sex

6.16f **Some crustaceans are able to change colour to avoid detection by predators. Colour change is achieved by the dispersal of pigments in specialised cells called**

    a. chromatocytes

    b. chromatophytes

c. chromatophils

d. chromatophores

6.17f The physiological problems experienced by desert animals have been extensively studied by

a. A. Macfadyen

b. E. P. Odum

c. K. Schmidt-Nielsen

d. D. H. Janzen

6.18f Where would you expect to find xerophytic plant species?

a. Hot deserts

b. Sand dunes in temperate climates

c. Strand lines on beaches

d. All of the above

6.19f Transpiration in flowering plants

a. increases in low wind speeds and stops completely in high wind speeds

b. stops at low wind speeds

c. increases at high wind speeds

d. is not affected by wind speed

6.20f A locust using incoming solar radiation from the sun to increase its body temperature when the sun is in the north-west is most likely to orientate the long access of its body

a. north-west – south-east

b. north – south

c. south-west – north-east

d. west – east

Intermediate

6.1i **In response to a recurring period of adverse environmental conditions some animal species (especially insects) exhibit a delay in their development that is known as**

a. hibernation

b. diapause

c. aestivation

d. acyclicity

6.2i **The plants of the sand dune grass species belonging to the genus *Ammophila* are able to roll up their leaves, with their stomata on the inside surface. This allows them to**

a. reduce their exposure to sunlight

b. reduce water loss via transpiration

c. increase their resistance to grazers

d. increase their photosynthetic rate

6.3i **Some animals hibernate to avoid adverse environmental conditions. Which of the following physiological changes is not characteristic of hibernation in black bears (*Ursus americanus*)?**

a. Increased metabolic rate

b. Reduced urine output

c. Lowered oxygen consumption

d. Lowered heart rate

6.4i **The source of the carbon used by plants in photosynthesis to make carbohydrates is**

a. atmospheric carbon monoxide

b. atmospheric carbon dioxide

c. calcium carbonate in the soil

b. carbon compounds in the soil water

6.5i The ability of an organism to time and repeat functions at intervals of approximately 24 hours, even when environmental clues such as the presence or absence of light are not available, is a manifestation of

    a. a tropism

    b. a circadian rhythm

    c. aestivation

    d. cyclicity

6.6i Birds such as petrels, gulls and fulmars spend much of their time in seawater and feeding on marine organisms. They take in excessive amounts of salt but are able to maintain osmotic balance by excreting sodium in their

    a. urine only

    b. nasal gland secretions only

    c. faeces and urine only

    d. nasal gland secretions and urine

6.7i The agama lizard (*Agama agama*) is a tropical species that is often found on the walls of houses. It uses different walls at different times of the day. This is most likely to be because

    a. it is a poikilotherm and moves onto the walls that receive most sun when it is cold and onto the shaded walls when it is too hot

    b. it is a homeotherm and moves onto the walls that receive most sun when it is cold and onto the shaded walls when it is too hot

    c. its eyes are too sensitive to tolerate bright light so it shelters on the shady walls when the sun is high in the sky and moves onto the sunny walls when the sun is low in the sky

    d. The movements of these lizards are entirely random and not driven by their physiological needs

6.8i Plants flower in response to changes in photoperiod, that is, changes in

    a. the relative lengths of the day and night at different times of the year

    b. the relative intensity of the light at different times of the year

c. the times of sunrise and sunset at different times of the year

d. the height of the sun in the sky at different times of the year

6.9i The water remaining in the soil when plants are in a state of permanent wilting due to water shortage, expressed as percentage dry weight of the soil, is called the

a. minimum soil capacity

b. drought point

c. compensation point

d. wilting point

6.10i Which types of savannah animals live in a structure constructed from soil in which the carbon dioxide concentration in the internal environment is controlled by a flow of air from the outside?

a. Ants

b. Bees

c. Termites

d. Dung beetles

6.11i Frost drought is most likely to affect

a. flowering plants

b. insects

c. birds

d. phytoplankton

6.12i Kangaroo rats (*Dipodomys* spp.) are desert rodents that are able to survive without access to free water in the environment. They obtain water primarily from the digestion of

a. seeds

b. insects

c. leaves

d. fruits

6.13i Bergmann's rule states that

    a. in poikilotherms, body size is larger in cooler regions and smaller in hotter regions

    b. in poikilotherms, body size is smaller in cooler regions and larger in hotter regions

    c. in homeotherms, body size is larger in cooler regions and smaller in hotter regions

    d. in homeotherms, body size is smaller in cooler regions and larger in hotter regions

6.14i Races of mammals and birds that live in cool, dry regions are lighter in colour (have less melanin pigment) than races of the same species that live in warm, humid regions. This is known as

    a. Gause's rule

    b. Gloger's rule

    c. Mayr's rule

    d. Simpson's rule

6.15i When a seal leaves the surface of the ocean and descends to great depth its physiological responses (e.g. changes in heart rate, the redistribution of blood, etc.) are referred to collectively as the

    a. diving adaptation

    b. submergence reflex

    c. diving reflex

    d. depth reflex

6.16i Plants tend to be more closely adapted to their environment than animals because

    a. plants evolved on Earth before animals

    b. animals can generally move away from adverse conditions but most plants are immobile

    c. animals are more tolerant of changes in environmental conditions than plants

    d. many plants live longer than most animals

6.17i Part of the surface of a flowering plant is shown in Fig. 6.1. The morphology of this plant suggests it is adapted to tolerate

    a. saltwater

    b. drought

    c. low temperatures

    d. low light intensity

Fig. 6.1.

6.18i A phreatophyte is a plant that derives its water from

    a. rainfall

    b. condensation

    c. groundwater

    d. metabolism

6.19i In a forest canopy the surface area of leaves per unit area of ground surface is the

    a. leaf area index

    b. canopy area index

    c. leaf: ground area index

    d. canopy area: ground area ratio

6.20i An organism that is adapted to taking advantage of unusual environmental conditions or newly created areas of habitat is known as

    a. a refugee species

    b. an escapee species

    c. a runaway species

    d. a fugitive species

Advanced

6.1a Which of the following lists most comprehensively represents the various methods used by different types of plants to obtain water?

    a. From capillary water in soil, groundwater, directly from the air, and from cloud and fog

    b. From capillary water in soil, groundwater, and from cloud and fog

    c. From capillary water in soil, groundwater and directly from the air

    d. From capillary water in soil, directly from the air, and from cloud and fog

6.2a Which of the following statements is false?

    a. Some fish species that inhabit Antarctic waters possess 'antifreeze proteins' which allow cells to survive in freezing conditions

    b. Ectotherms need more oxygen than endotherms of the same size

    c. Endotherms are also known as homeotherms

    d. Some endotherms have a lower internal body temperature than some ectotherms

6.3a The sigmoid curve (S) in Fig. 6.2 is the oxygen–haemoglobin dissociation curve for a deer mouse (*Peromuscus maniculatus*) living at low altitude. Where on this graph would you expect the line to appear for a deer mouse adapted to living at a higher altitude and what form would you expect the line to take?

a. A sigmoid curve located to the right of S

b. A sigmoid curve located to the left of S

c. An exponential curve joining points t and u

d. A straight line joining points t and u

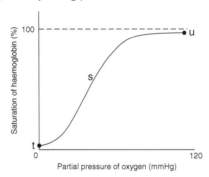

Fig. 6.2.

6.4a Camels store excess heat in their bodies when they are dehydrated and deprived of water. The rectal temperatures of four species of mammals (A–D) are shown in Table 6.1. Which is a camel?

a. A

b. B

c. C

d. D

Table 6.1

Species	Rectal temperature	
	Mean (°C)	Range (°C)
A	39.1	38.5–39.7
B	39.5	38.6–40.1
C	37.6	37.2–38.1
D	37.5	34.2–40.7

93

6.5a In marine teleost fishes specialised cells, known as chloride cells, assist in osmoregulation by removing NaCl from plasma to seawater by active transport. These cells are located in the

    a. kidneys

    b. gills

    c. rectal gland

    d. oral cavity

6.6a Sharks are able to maintain elevated temperatures in some regions of the body, for example the muscles used for swimming and parts of the digestive tract. They are able to do this because

    a. they are homeothermic

    b. they consume large quantities of high-energy food

    c. their bodies contain heat exchangers that transfer heat from the gills to the tissues to be heated

    d. they transfer metabolic heat from the liver to the tissues that are heated

6.7a Some lizards are nocturnal so cannot absorb solar radiation directly. Instead they increase their body temperature by the conduction of heat from warm rocks and sand. These animals are known as

    a. thigmotherms

    b. heliotherms

    c. thermosaurs

    d. endotherms

6.8a The equation below is used to calculate

    a. the heat loss from an organism's body due to respiration

    b. the convectional exchange of energy between an organism and the air

    c. the heat gain experienced by an organism due to solar radiation

    d. the heat transferred from an organism's body to the ground by conduction

$$C = h_c(T_0 - T_a)$$

Where T_0 = temperature of the organism's surface, T_a = air temperature, and h_c = a coefficient that includes the wind velocity and the size of the organism.

6.9a **The densities of stomata on the upper and lower leaf surfaces of four plant species (A–D) are shown in Table 6.2. Identify the plant species that is a monocotyledon.**

    a. A

    b. B

    c. C

    d. D

Table 6.2

Plant species	Number of stomata on the leaf surface/ cm²	
	Upper epidermis	Lower epidermis
A	5200	6800
B	4000	28100
C	0	45000
D	1200	13000

6.10a **Why do thermogenic plants (e.g. the skunk cabbage (*Symplocarpus foetidus*) found in the swamps of the northern United States) raise the temperature of their flower parts to 10–15°C above that of the environment?**

    a. To reduce the amount of energy required for photosynthesis

    b. To reduce their susceptibility to frost

    c. To remove excess energy produced by metabolism

    d. To volatilize chemicals that attract pollinating insects

6.11a **Subnivean animals live under**

    a. rocks

    b. snow

c. tree roots

d. the seabed

6.12a **A chionophile is an organism that is adapted to life in an environment that is**

a. cold

b. dry

c. hot

d. humid

6.13a **Twisted, scrubby or depressed 'trees' (krummholz) found at arctic and alpine tree lines are the result of**

a. wind action

b. low temperatures during the growing season

c. damage from snow and ice particles

d. all of the above

6.14a **The alpine sorrel (*Oxyria digyna*) is a widely distributed plant that occurs in the Arctic and high mountain regions of North America and Eurasia. It grows primarily where snow does not melt until midsummer and its distribution may be described as**

a. circumpolar

b. circumboreal

c. boreal

d. alpine

6.15a **A plant experiencing the light level and carbon dioxide concentration indicated by point A in Fig. 6.3 would increase its rate of photosynthesis if**

a. light level alone increased

b. carbon dioxide concentration alone increased

c. either light level or carbon dioxide concentration increased

d. light level increased and carbon dioxide level decreased

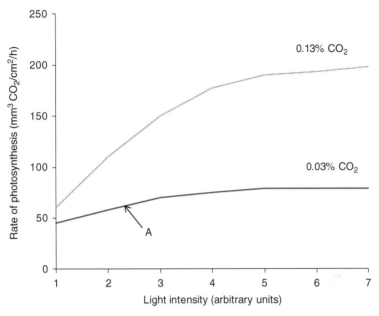

Fig. 6.3.

6.16a Changes in the rate of oxygen consumption of a large mammal that hibernates during part of the year are shown in Fig. 6.4.

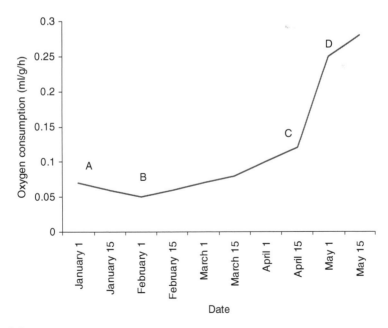

Fig. 6.4.

Which of the points A to D indicates the point in time when the animal emerged from hibernation?

a. A

b. B

c. C

d. D

6.17a The responses of four species of midges (A–D) to a lack of oxygen in their aquatic environment are shown in Fig. 6.5. Which species is best adapted to survive in water that has been heavily polluted with organic matter?

a. A

b. B

c. C

d. D

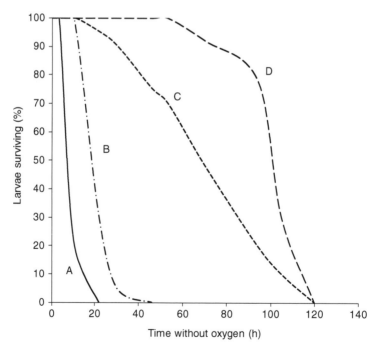

Fig. 6.5.

6.18a **In which regions of the visible spectrum does chlorophyll absorb light?**

    a. Red and blue

    b. Red and green

    c. Blue and green

    d. Red only

6.19a **A calcifuge is a plant associated with**

    a. an acidic soil

    b. an alkaline soil

    c. a waterlogged soil

    d. a desert soil

6.20a **The relationship between photosynthesis (measured as carbon dioxide consumed) and cellular respiration (measured as carbon dioxide released) in a plant at different light intensities is shown in Fig. 6.6. Which point on the curve is the compensation point?**

    a. A

    b. B

    c. C

    d. D

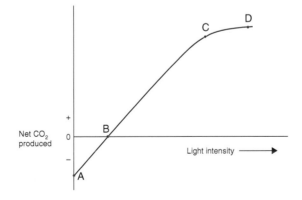

Fig. 6.6.

7 Population Ecology

This chapter contains questions about the growth, control, regulation and analysis of biological populations.

Foundation

7.1f **Which of the following could not be described as a population?**

    a. All of the Asian elephants in Yala National Park, Sri Lanka

    b. All of the reptiles in Death Valley, California

    c. All of the killer whales in the ocean

    d. All of the red kites in Wales

7.2f **The natality rate of a population is the same as its**

    a. death rate

    b. birth rate

    c. morbidity rate

    d. emigration rate

7.3f **A sudden increase in the numbers of an insect species (e.g. a locust) in response to a change in the environment is known as**

    a. an outbreak

    b. an invasion

c. an irruption

d. an expansion

7.4f The geographical area where a population of animals lives is called its

a. range

b. territory

c. habitat

d. environment

7.5f Which of the following is not a characteristic of a population?

a. density

b. natality

c. mortality

d. genotype

7.6f The number of plants growing in an area divided by the size of the area gives its

a. density

b. population size

c. dispersion

d. intensity

7.7f Population size is a function of

a. birth rate and death rate

b. birth rate and immigration rate

c. birth rate, death rate and emigration rate

d. birth rate, death rate, immigration rate and emigration rate

7.8f Which of the following is not a density-dependent factor which could influence the size of a population of herbivorous insects?

a. an extended period of frost

b. inter-specific competition

c. a viral disease

d. predation by birds

7.9f **The number of individuals in each age class of a population of mammals is shown in Fig. 7.1. In the immediate future this population is likely to**

a. increase in size

b. decrease in size

c. remain stable

d. become extinct

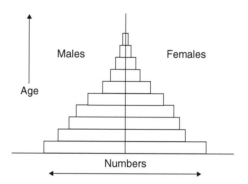

Fig. 7.1.

7.10f **A cohort is a group of individuals of the same**

a. age

b. sex

c. size

d. genotype

7.11f **The simplest index of mortality in a population is the**

a. age-specific death rate

b. instantaneous death rate

c. finite death rate

d. crude death rate

7.12f **A species that is semelparous**

    a. reproduces many times during its life

    b. reproduces once during its life

    c. has more than one offspring at a time

    d. may be caring for young of more than one age

7.13f **Red deer (*Cervus elaphus*) hinds do not reproduce for the first two years of life. In which stage of her life is a hind that is 18 months old?**

    a. Post-reproductive

    b. Pre-reproductive

    c. Mid-reproductive

    d. Peri-reproductive

7.14f **All of the okapis (*Okapia johnstoni*) kept in zoos that collaborate in a captive-breeding programme by exchanging animals may be considered to be a**

    a. subpopulation

    b. megapopulation

    c. metapopulation

    d. micropopulation

7.15f **Which of the following statements about populations is false?**

    a. All of the individuals in a population are capable of interbreeding

    b. Populations increase as a result of births and immigration

    c. Populations decrease as a result of deaths and emigration

    d. A population may consist of individuals from more than one species

7.16f **A population that cannot survive without immigrants from nearby populations is called a**

    a. sink population

    b. supported population

    c. source population

    d. founder population

7.17f The study of populations is called

    a. dendrology

    b. demography

    c. deontology

    d. desmology

7.18f A female mammal that has not produced any offspring may be described as

    a. infertile

    b. impotent

    c. anoestrous

    d. nulliparous

7.19f The cyclic seasonal movement of a population of animals from one geographical location to another is a

    a. translocation

    b. migration

    c. displacement

    d. relocation

7.20f Which of the following mammal species is well known for its population cycles?

    a. Norwegian lemming (*Lemmus lemmus*)

    b. Black-footed ferret (*Mustela nigripes*)

    c. European badger (*Meles meles*)

    d. Pine marten (*Martes martes*)

Intermediate

7.1i A table showing the death rates of different age classes of individuals in a population of brown bears (*Ursus arctos*) is called

    a. a death table

    b. a morbidity table

c. a life table

d. a survival table

7.2i **The growth of a population is illustrated in Fig. 7.2. Which section of the curve indicates the asymptote?**

a. 1

b. 2

c. 3

d. 4

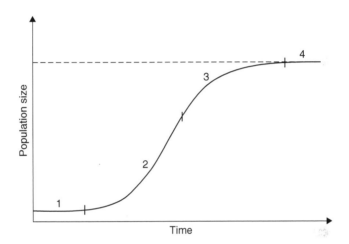

Population size

Time

Fig. 7.2.

7.3i **The Lincoln index is used to estimate the size of a population of mobile animals by catching a sample, marking them and releasing them into the population, then a short time later catching a second sample and determining what proportion of the second sample carries marks. Which of the following assumptions are made by the Lincoln index? Between the first trapping (when animals are marked) and the second trapping**

(i) no immigration occurs

(ii) no emigration occurs

(iii) no births occur

(iv) no deaths occur

(v) no marks are lost

a. i, ii and iii only

b. i, ii, iii, iv and v

c. ii, iv and v only

d. v only

7.4i **Which of the following sequences of numbers represents exponential growth? (Each sequence represents the size of a population at fixed time intervals).**

a. 3, 5, 8, 16, 22, 64, 129, 125, 129, 127, 130

b. 17, 25, 36, 53, 109, 136, 227, 401, 732

c. 2, 4, 9, 15, 32, 64, 127, 92, 75, 38, 21

d. 2, 4, 8, 16, 32, 64, 128, 256, 512, 1024

7.5i **The data in Table 7.1 relate to a population of animals. In 2020 the natality rate of the population was**

a. 162/579 per year

b. 93/579 per year

c. (162–93)/579 per year

d. 162/648 per year

Table 7.1

Population size at beginning of 2020	579
Births during 2020	162
Deaths during 2020	93
Population size at the end of 2020	648

7.6i **Name the types of population growth shown in graphs A, B and C in Fig. 7.3 by matching them with the appropriate section of Table 7.2**

a. 1

b. 2

c. 3

d. 4

Table 7.2

	1	2	3	4
A	Boom-and-bust	Logistic	Logistic	Exponential
B	Logistic	Boom-and-bust	Exponential	Logistic
C	Exponential	Exponential	Boom-and-bust	Boom-and-bust

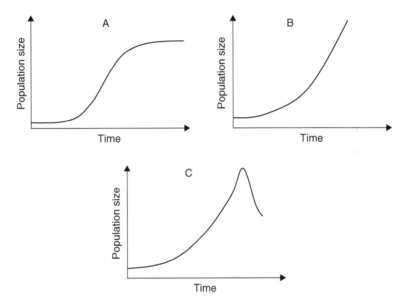

Fig. 7.3.

7.7i How many rapid rises (pulses) in population growth are seen in the North Atlantic Ocean diatom population in each year?

    a. One, in summer

    b. One in spring and one in summer

    c. One in spring and one in autumn

    d. None

7.8i When a species is introduced into an area which it has not occupied previously, where the environment is favourable and contains no predators or competitors, its initial population growth is likely to be

    a. stable

    b. exponential

c. logistic

d. J-shaped

7.9i   To construct a static life table an ecologist should determine

a. the age of all of the animals present in a population at a particular point in time

b. the age at which each animal dies from a cohort of individuals that were all born at the same time (e.g. in the same year)

c. the number of offspring produced in a population each year

d. the number of deaths that occur in the population each year

7.10i   In a life table, the abbreviation q_x refers to

a. the number of births in an age class

b. the number of survivors in an age class

c. the mean expectation of further life in an age class

d. the age-specific death rate

7.11i   Which of the survivorship curves shown in Fig. 7.4 is typical of an animal parasite?

a. 1

b. 2

c. 3

d. A line midway between 1 and 2

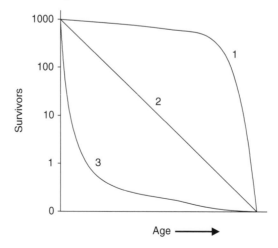

Fig. 7.4.

7.12i **Changes in the size of an animal population are shown over 24 years in Fig. 7.5. The arrow indicates a point where**

a. the carrying capacity of the environment decreased

b. the carrying capacity of the environment increased

c. the growth rate of the population increased

d. the mortality rate of the population decreased

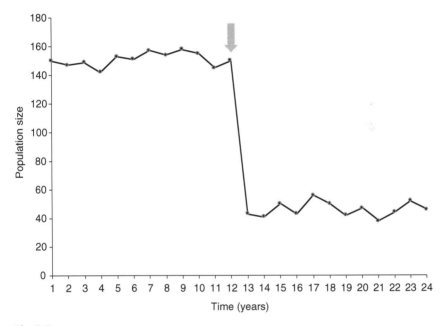

Fig. 7.5.

7.13i **An ecologist followed the fate of all the female antelope born into one population in 1995. She recorded the date each individual died and created a table showing the number of individuals alive in each year until all the animals had died. To study survivorship in this population from these data she constructed a**

a. dynamic life table

b. a stationary life table

c. a static life table

d. a mobile life table

7.14i The Lincoln index was used to estimate the size of a population of squirrels. What is the estimate of the population size if, 20 squirrels were captured during the first trapping session, marked and released, 12 were captured during the second trapping session and 6 of these were marked?

a. 4

b. 5

c. 10

d. 40

7.15i The changes in the population size of the mule deer (*Odocoileus hemionus*) on the Kaibab Plateau in Arizona in the early part of the 20th century have often been cited as an example of

a. logistic population growth

b. exponential population growth

c. boom-and-bust population growth

d. stable population growth

7.16i Which of the following statements about survivorship curves is false?

a. They are usually based on a starting population of 1000 individuals

b. Males and females are normally shown as separate lines

c. Age is plotted against the vertical (y) axis

d. The curve is derived from data in a life table

7.17i Which of the following characteristics of a species is most difficult to define?

a. rarity

b. density

c. natality rate

d. mortality rate

7.18i The size of a population of squirrels was estimated using the Lincoln index. On the first trapping occasion 25 animals were captured, marked with collars and released. On the second trapping occasion 18 squirrels were captured and 10 had collars. Unknown to the researchers, 3 of the marked squirrels had removed their collars prior to the second trapping occasion. What would be the likely effect of this loss of marks on the population estimate?

a. The estimate would not be affected

b. The estimate would be too high

c. The estimate would be twice what it should be

d. The estimate would be too low

7.19i Which of the following statements about a life table is false?

a. It can be used to produce a survivorship curve

b. It has a column representing age-specific death rates

c. A static life table represents a population at a single point in time

d. It only ever shows data for females

7.20i Some methods of estimating the size of a population of small mammals require them to be trapped and marked on two or more occasions. Such methods assume that all individuals in the population are equally likely to be caught (whether marked or not) but some animals, having been caught once, will not enter a trap again. These individuals are

a. trap happy

b. trap shy

c. trap intolerant

d. trap reticent

Advanced

7.1a Undercrowding (lack of aggregation) may limit population growth; as the population size increases survival and reproduction increase. This is known as

a. Allen's principle

b. Allee's principle

c. Andrewartha's principle

d. Ayala's principle

7.2a **The differential equation below is used to describe a type of population growth. What is r?**

$$\frac{dN}{dt} = rN$$

a. Intrinsic rate of natural increase

b. Asymptote

c. Initial size of the population

d. Death rate

7.3a **The differential equation below describes the continuous version of the logistic model of population growth.**

$$\frac{dN}{dt} = \frac{rN(K-N)}{K}$$

In this equation, the asymptote (carrying capacity) is denoted by

a. r

b. K

c. dN

d. rN

7.4a **The growth of two populations of the same species (A and B) is shown in Fig. 7.6. Which of the following variables is higher in population A than in population B?**

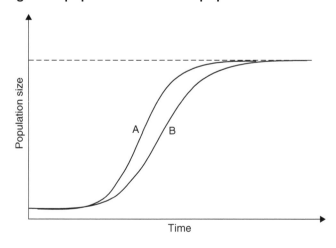

Fig. 7.6.

112

a. *K*

b. *r*

c. The initial value of *N*

d. *t*

7.5a **Population growth may be studied using a Leslie matrix. In the matrix below, the likelihood of an individual surviving from year 2 to year 3 is given by**

$$\begin{bmatrix} n_{0t+1} \\ n_{1t+1} \\ n_{2t+1} \end{bmatrix} = \begin{bmatrix} f_0 & f_1 & f_2 \\ p_0 & 0 & 0 \\ 0 & p_1 & 0 \end{bmatrix} \times \begin{bmatrix} n_{0t} \\ n_{1t} \\ n_{2t} \end{bmatrix}$$

a. p_1

b. n_{1t}

c. p_0

d. f_1

7.6a **An estimate of the size of a fish population calculated by plotting the number of fish caught in individual consecutive catches against the cumulative total catch on days 1 to 8 is shown in Fig 7.7. The method used was**

a. mark-recapture

b. total count

c. stratified sampling

d. removal trapping

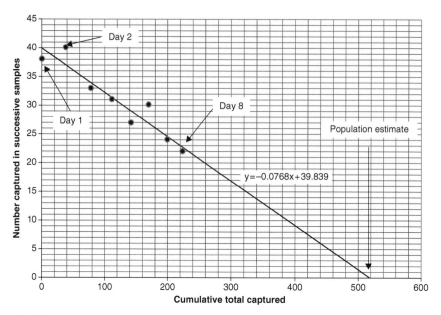

Fig. 7.7.

7.7a The effect of a parasitic insect (*Eurytoma curta*) on the population of its host, the knapweed gallfly (*Urophora jaceana*), is shown in Table 7.3. Which of the following statements is supported by these data?

a. This parasite has a density-independent effect on its host

b. Insect parasites have a density-dependent effect on their hosts

c. This parasite has a density-dependent effect on its host

d. As the density of gallfly larvae increases the percentage killed by *Eurytoma curta* remains constant

Table 7.3

Year	Knapweed gallfly larval population/m² at the beginning of the season	Larvae killed by parasite (*Eurytoma curta*)/m²
1934	43	6
1935	148	66

7.8a Table 7.4 shows the percentage of adult female African elephants (*Loxodonta africana*) found to be pregnant during culling operations in four locations in Uganda (A–D) by Laws *et al.* (1975) and the mean calving interval (years) in these populations. Which of the following statements is not supported by these data?

    a. If calving interval increases birth rate increases

    b. Increasing elephant population density may be reducing the frequency with which individual elephants produce calves

    c. Habitat deterioration may be causing a reduction in birth rate

    d. The combined effect of habitat deterioration and increasing elephant population density may be reducing elephant population growth

Table 7.4

Habitat deterioration	Population density	Population	Adult females pregnant (%)	Mean calving interval (years)
Lowest	Lowest	A	63.33±9.95	2.9
Low	Low	B	43.33±10.24	4.2
High	High	C	26.88±9.02	6.8
Highest	Highest	D	20.17±4.22	9.1

7.9a Which of the following could be described is an *r*-strategist?

    a. Locust (*Locusta migratoria*)

    b. *Tyrannosaurus rex*

    c. Blue whale (*Balaenoptera musculus*)

    d. *Homo sapiens*

7.10a Which of the following could be described as a *K*-strategist?

    a. A bacterium

    b. An aphid

    c. An elephant

    d. A virus

7.11a The Lincoln index is used to estimate the size of a population of mobile organisms by catching individuals on one occasion (number caught = n_1), marking them, releasing them
and then catching a second sample (number caught = n_2).
The number within the second sample that bear marks (i.e.
those caught on both occasions) = r. Which of the following
formulae represents the Lincoln index?

a. $\dfrac{r}{n_1 \times n_2}$

b. $\dfrac{n_1 \times n_2}{r}$

c. $\dfrac{n_1}{n_1 \times r}$

d. $\dfrac{r \times n_2}{n_1}$

7.12a The morbidity rate of a population is a measure of its

a. rate of incidence of disease

b. death rate

c. fecundity

d. peri-natal death rate

7.13a Female wildebeest (*Connochaetes taurinus*) in the Serengeti
National Park time the births of their calves so that they are
all produced around the same time as a means of reducing
losses due to predation by lions and other carnivores. This
is achieved by

a. oestrus synchrony

b. social facilitation

c. mating co-ordination

d. anoestrus synchrony

7.14a The growth of an animal population when it is unrestricted
(1) and when it is constrained by factors in the environment
(2) is shown in Fig. 7.8. Which of the areas P, Q, R or S represents 'environmental resistance'?

a. P

b. Q

c. R

d. S

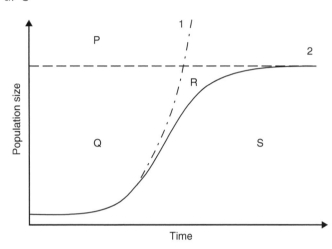

Fig. 7.8.

7.15a The visual and auditory stimulation that is experienced by some large breeding colonies of birds (e.g. gulls) causes an acceleration and synchronisation of breeding resulting in a reduction in the likelihood that any particular pair will have their eggs taken by predators. This is known as the

a. Fraser Smith effect

b. Elliot Darling effect

c. Charles Darwin effect

d. Fraser Darling effect

7.16a In a population of higher plants, if N_t = the number of adult plants in the population; F = the average fecundity; g = the proportion of seeds that germinate on average; and e is the probability of an individual seedling establishing itself as an independently photosynthesising adult, how would you calculate the total number of new adult plants produced?

    a. $N_t \times F \times g \times e$

    b. $(N_t \times F \times g)/e$

    c. $N_t + F + g + e$

    d. $N_t \times F - g \times e$

7.17a Population *regulation* by definition can only occur as a result of

    a. density-dependent or density-independent processes

    b. density-independent processes alone

    c. density-dependent processes alone

    d. physical factors in the environment

7.18a The population shown in Fig. 7.9 is exhibiting a population fluctuation known as

    a. chaos

    b. stable limit cycles

    c. damped oscillations

    d. monotonic damping

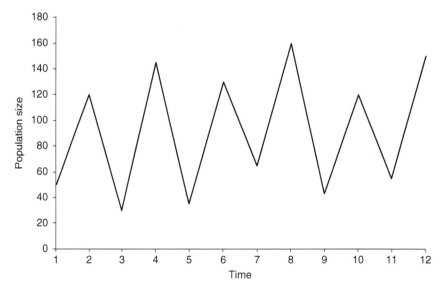

Fig. 7.9.

7.19a Age-specific fecundity for an animal population is shown in Fig. 7.10. What does axis Y represent?

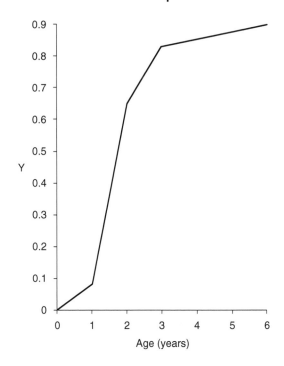

Fig. 7.10.

    a. Number of survivors per individual

    b. Population size

    c. Total number of births in the population

    d. Number of births per individual

7.20a Consider the following models of population growth:

Model 1	$\dfrac{dy}{dt} = ay - by^2$
Model 2	$\dfrac{dy}{dt} = ry$
Model 3	$\dfrac{dy}{dt} = \left[a + y(t)\right]y$

Where,

t = time

y = population density at time t

r = a constant

a = a constant

b = a constant

$y(t)$ = a random variable with mean zero

Which of the following statements about these models is true?

a. All of the models are deterministic

b. Models 1 and 2 are deterministic and model 3 is stochastic

c. Models 1 and 2 are stochastic and model 3 is deterministic

d. Models 2 and 3 are stochastic and model 1 is deterministic

8 Community Ecology and Species Interactions

This chapter contains questions about the structure of biological communities, their development over time, and the interactions between the species living within them, especially competition, niche separation and predation.

Foundation

8.1f The existence of distinct vertical layers of vegetation within a community, each with its own assemblage of plants, is known as

    a. zonation

    b. stratification

    c. succession

    d. lamination

8.2f An animal that transmits a disease or parasite from one organism to another is called

    a. a zoonosis

    b. an agent

    c. a prion

    d. a vector

8.3f   African elephants (*Loxodonta africana*) have a dispropor-
tionately large effect on the community in which they occur,
converting forests to grassland, and creating waterholes
from which other species also benefit. For this reason
elephants are referred to as

    a. an allopatric species

    b. a pioneer species

    c. a megaherbivore

    d. a keystone species

8.4f   Which of the following pairs of animals was the subject
of a famous predator-prey study conducted by Elton and
Nicholson?

    a. Rabbit and fox

    b. Hare and lynx

    c. Rabbit and bobcat

    d. Hare and puma

8.5f   In which of the following countries would you find cloud
forest?

    a. England

    b. Canada

    c. Costa Rica

    d. Niger

8.6f   Taiga is another term for

    a. tropical forest

    b. temperate forest

    c. boreal forest

    d. mangrove forest

8.7f   A riparian forest is a forest located

    a. on the side of a mountain

    b. adjacent to a river or stream

c. in an urban area

d. at the edge of a desert

8.8f **The term 'benthos' refers to the community of organisms that lives**

a. at the bottom of a lake, river or sea

b. on the floor of a tropical forest

c. beneath the leaf litter in a temperate woodland

d. in the soil in tundra

8.9f **An animal parasite living on the outside of the body of another animal is called**

a. an ectoparasite

b. an endoparasite

c. an alloparasite

d. a sero-parasite

8.10f **Algal species occur in bands on an exposed rocky shore. These bands run more or less parallel to the shoreline and the location of each species depends on the amount of time it can survive exposure to the air when the tide recedes. This pattern of distribution is known as**

a. zonation

b. stratification

c. lamination

d. banding

8.11f **The final stage of succession, when the vegetation has developed to a certain stage of equilibrium, is called the**

a. pioneer community

b. steady state community

c. equilibrium community

d. climax community

8.12f The organisms growing on the surface of the tree in Fig. 8.1 are called

a. exophytes

b. epiphytes

c. xerophytes

d. halophytes

Fig. 8.1.

8.13f Tundra, hot desert, temperate grassland and taiga are all examples of

a. biomes

b. seres

c. ecotypes

d. ecotones

8.14f **The guts of ruminants contain bacteria and protozoa that digest cellulose and lignin into compounds that the host animal can use. This association is known as**

    a. herbivory

    b. mutualism

    c. parasitism

    d. assimilation

8.15f **Competition between two species for the same resource is known as**

    a. interspecific competition

    b. intraspecific competition

    c. infraspecific competition

    d. hyperspecific competition

8.16f **Which of the following attributes is possessed by a community but not by a population?**

    a. Birth rate

    b. Density

    c. Death rate

    d. Species diversity

8.17f **A seral stage is a step in the process of**

    a. photosynthesis

    b. ecological succession

    c. decomposition

    d. tree growth

8.18f **The habitat shown in Fig. 8.2 is**

    a. mangrove forest

    b. boreal forest

    c. cloud forest

    d. temperate forest

Fig. 8.2.

8.19f **The main stabilising agent in sand dune ecosystems in the United Kingdom is**

    a. marram grass (*Ammophila* spp.)

    b. couch grass (*Elymus repens*)

    c. sea bindweed (*Calystegia soldanella*)

    d. coniferous forest

8.20f **In boreal forests, lakes gradually fill with organic matter produced by sphagnum moss (*Sphagnum* spp.) or sedges, and may become completely covered with a floating un-stable mat of vegetation forming a**

    a. marsh

    b. swamp

    c. mere

    d. peat bog

Intermediate

8.1i The Pantanal is a tropical wetland of high biodiversity located in

    a. Africa

    b. India

    c. Indonesia

    d. South America

8.2i The niches of insect species in two hypothetical ecosystems are represented by diagrams A and B in Fig 8.3. These ecosystems are

    a. A: tundra; B: temperate deciduous forest

    b. A: savannah; B tropical rainforest

    c. A: boreal forest; B: tropical rainforest

    d. A: tropical rainforest; B: tundra

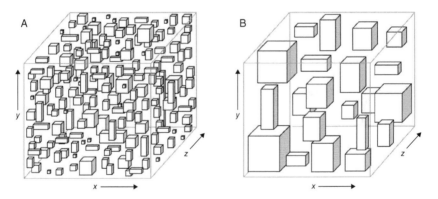

Fig. 8.3.

8.3i Similar species occupying overlapping ranges may engage in resource partitioning to avoid

    a. predation

    b. interspecific competition

    c. intraspecific competition

    d. parasitism

8.4i Gause's law states that 'complete competitors cannot coexist'. It is also known as the

   a. competitive coexistence principle

   b. coexistence and competition theory

   c. competitive exclusion principle

   d. competitive expulsion theory

8.5i The use of a natural parasite or predator to control the numbers of a pest species is most commonly described as

   a ecological control

   b. biological control

   c. pest control

   d. organic control

8.6i The graph below (Fig. 8.4) illustrates

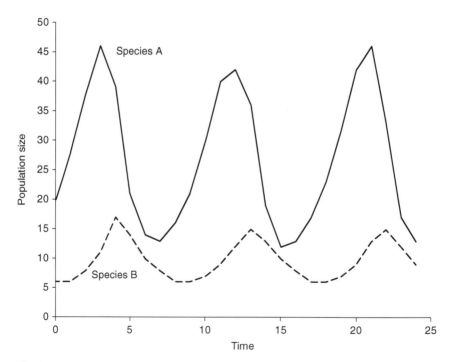

Fig. 8.4.

   a. predator–prey interactions

   b. interspecific competition

   c. intraspecific competition

   d. logistic population growth

8.7i **A guild is**

   a. a group of prey species fed upon by the same predator

   b. a group of colonial insects

   c. a group of species that exploit the same resources

   d. an alternative name for a niche

8.8i **When a palaeoecologist reconstructs past communities of flowering plant species that lived in a particular area from soil cores, the most appropriate method involves**

   a. leaf analysis

   b. seed analysis

   c. pollen analysis

   d. root analysis

8.9i **Which of the following statements about niches is false?**

   a. If it is introduced into an ecosystem with a suitable vacant niche an invasive species may expand rapidly

   b. Niche breadth in tropical rainforest species is generally wide

   c. The fundamental niche of a species is larger than its realised niche

   d. The term 'niche partitioning' means the same as 'niche separation'

8.10i **The palaeoecology of past mountains ranges is little understood because**

   a. it is in the nature of mountains that they become eroded with time so are unsuitable environments for discovering fossils

   b. mountains support relatively few species so their fossils are less likely to be found

c. climatic conditions in mountainous areas do not favour fossilisation

d. mountain species tend to be small and therefore their fossils do not survive

8.11i Malaria is a disease transmitted by

a. *Anopheles* acting as a vector for *Plasmodium* species

b. *Glossina* acting as a vector for *Trypanosoma* species

c. *Anopheles* acting as a vector for *Trypanosoma* species

d. *Glossina* acting as a vector for *Plasmodium* species

8.12i Following the abandonment of agricultural fields in the southeastern Unites States, a study by Johnston and Odum (1956) recorded the sequence of events shown in Table 8.1.

Table 8.1

Dominant plants	Age of study area (years)	Pairs of birds/100 acres	Number of bird species
Forbs	1–2	15	2
Grasses	2–3	40	2
Grass-shrub	15	110	7
	20	136	9
Pine forest	25	87	7
	35	93	10
	60	158	20
	100	239	18
Oak-hickory forest	150–200	228	19

This is an example of

a. zonation

b. a primary succession

c. a diversity gradient

d. a secondary succession

8.13i **Mature plants of the same species growing in close proximity may compete with each other for resources. This may result in which of the following?**

a. Density-dependent mortality of plants

b. Reduced seed production

c. A decrease in the number of vegetative offspring produced

d. All of the above are possible responses to this type of competition

8.14i **Which of the following is not a trend observed during a primary succession on newly exposed rock?**

a. Vegetation becomes taller

b. Soil becomes deeper

c. Insect species diversity decreases

d. Soil organic matter depth increases

8.15i **In 1883 the volcanic island of Krakatau exploded, leaving a lifeless surface. Twenty-five years later the island was covered in dense forest and some 263 animal species. This illustrates the phenomenon of**

a. reintroduction

b. dispersal

c. dispersion

d. evolution

8.16i **Which biome has the following characteristics?**

Evergreen conifers dominate, continuous stands, no lower layers of vegetation, relatively few species, slow decay of litter, ground layer of lichens and bog moss.

a. Tropical rainforest

b. Temperate deciduous forest

c. Boreal forest (taiga)

d. Mangrove forest

8.17i Which of the following statements is false?

    a. In commensalism, one organism benefits from the relationship, the other is more or less unharmed

    b. In mutualism both organisms benefit

    c. In parasitism one organism benefits, the other is harmed

    d. In interspecific competition one species always eventually becomes extinct

8.18i The phytophysiognomy of a plant community is concerned with its

    a. physical structure

    b. species composition

    c. water tolerance

    d. spatial distribution

8.19i An ecologist ascended the Ruwenzori Massif in East Africa. Which of the following sequences accurately represents the altitudinal zones she passed through while climbing from the bottom to the top?

    a. rainforest, savannah, bamboo, heath, alpine

    b. savannah, rainforest, bamboo, heath, alpine

    c. savannah, rainforest, bamboo, alpine, heath

    d. heath, savannah, rainforest, bamboo, alpine

8.20i Which biome has the following characteristics?

Marked seasonality of climate, large herbivorous mammals, maintained by fire, associated with plains and plateaux, tall grasses, only two vegetation layers.

    a. Temperate grassland

    b. Tropical savannah

    c. Tidal marshland

    d. Cloud forest

Advanced

8.1a **'Environmental resilience' is usually taken to be a characteristic of**

a. species

b. populations

c. ecosystems

d. food chains

8.2a **A classic study of predation by Thomas Paine demonstrated that**

a. a starfish (*Pisaster*) could simultaneously regulate the numbers of several of its prey species

b. predation by lions (*Panthera leo*) could regulate the numbers of wildebeest (*Connochaetes taurinus*) in the Serengeti

c. the abundance of several freshwater species in a river system was determined by predation pressure from otters (*Lutra lutra*)

d. predation by bald eagles (*Haliaeetus leucocephalus*) regulates the size of black-footed ferret (*Mustela nigripes*) populations

8.3a **The niches occupied by species A and B with respect to three variables (X, Y, and Z) are illustrated in Fig. 8.5. Which of the following answers to the question 'Do their niches overlap?' is false?**

a. No, if all three variables are considered

b. Yes, if only variable Z is considered

c. No, if only variables X and Y are considered

d. Yes, if only Y and Z are considered

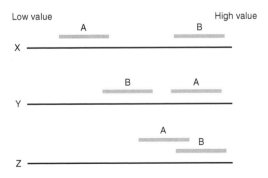

Fig. 8.5.

8.4a **The locations (heights) where individuals from three species of warblers (A, B and C) were observed feeding in a coniferous forest are indicated in Fig. 8.6. This is preliminary evidence of**

a. intraspecific competition

b. niche partitioning

c. competitive exclusion

d. a species diversity gradient

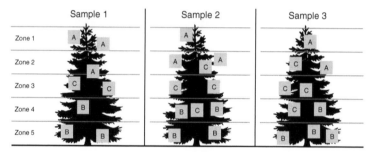

Fig. 8.6.

8.5a **A field experiment that was set up in a desert area of North America by ecologists interested in the interaction between rodents and ants is illustrated in Fig. 8.7. Both taxa used seeds as food. Four enclosures were established. In enclosure A all rodents were removed; in enclosure B all ants were removed; in enclosure C both taxa were removed; and in enclosure D (the control) both taxa were present. The seed density was deemed to be 1.0 in the control enclosure (D) where no animals had been removed. Which of the following statements is false?**

a. The number of ant colonies was highest when rodents were absent

b. The number of rodents was highest when ant colonies were absent

c. When ant colonies and rodents were present the numbers of both were suppressed

d. When neither ant colonies nor rodents were present the seed density was 10 times higher than when both were present

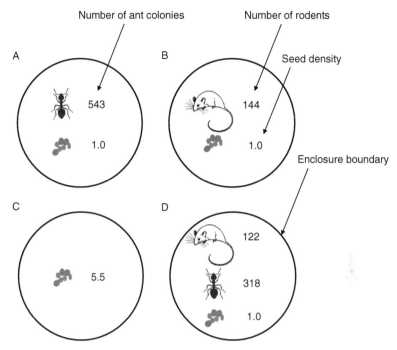

Fig. 8.7.

8.6a **Which of the following statements about niches is false?**

a. Charles Elton classified niches based on the feeding habits of organisms

b. The first biologist to describe the niche was Ernst Haeckel

c. McArthur and Levins discussed the niche in terms of resource utilisation

d. Grinnell defined an organism's niche in terms of the habitat it occupies

8.7a **The concept of the niche proposed by Hutchinson may be described in terms of**

a. a 3-dimensional hyperspace

b. an n-dimensional space

c. an n-dimensional hypervolume

d. a 2-dimensional megavolume

8.8a **The competition between two species (1 and 2) for food may be described by the Lotka-Volterra equations given below.**

Population growth of species 1 in competition:

$$\frac{dN_1}{dt} = r_1 N_1 \left(\frac{K_1 - N_1 - \alpha N_2}{K_1} \right)$$

Population growth of species 2 in competition:

$$\frac{dN_2}{dt} = r_2 N_2 \left(\frac{K_2 - N_2 - \beta N_1}{K_2} \right)$$

In these equations, the term βN_1 represents

a. the effect of species 1 on the growth of species 2

b. the intrinsic rate of increase of species 1

c. the effect of species 2 on the growth of species 1

d. the intrinsic rate of increase of species 2

8.9a **Parasitism is sometimes considered to be a special kind of**

a. herbivory

b. competition

c. predation

d. symbiosis

8.10a **A plagioclimax is a community that**

a. has developed on sandy soil

b. is maintained by continuous human activity

c. always contains large tree species

d. can only occur in a temperate climate

8.11a *The Theory of Island Biogeography* **was written by**

a. Lee and Gates

b. Grieve and Went

c. Barlett and Gates

d. MacArthur and Wilson

8.12a An archipelago of volcanic islands 10km from the mainland is illustrated in Fig. 8.8. The numbers indicate the number of species of small seed-eating birds that have colonised each island. In which direction is the mainland most likely to be?

    a. North

    b. South

    c. West

    d. East

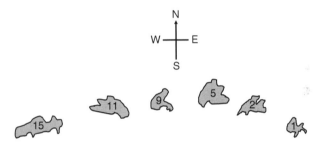

Fig. 8.8.

8.13a Who performed laboratory experiments in the 1930s to test the Lotka-Volterra equations?

    a. G. F. Gause

    b. E. Hutchinson

    c. E. P. Odum

    d. R. MacArthur

8.14a The American ecologist Robert MacArthur conducted a famous study of

    a. predator–prey interactions in mammals

    b. niche separation in warblers

    c. species diversity gradients in desert lizards

    d. intraspecific competition in plants

8.15a A prey organism that can run quickly from its predator and a predator that can run quickly after its prey are both favoured by natural selection. This interaction between predators and their prey is known as

a. an evolutionarily stable strategy

b. resource partitioning

c. adaptive variation

d. an evolutionary arms race

8.16a Prey switching – also called predator switching – is the phenomenon whereby

a. predators change their food preference to whichever suitable prey is most common

b. predators change their method of hunting depending upon the size of the prey animal

c. predators switch to nocturnal hunting to avoid daytime disturbance by tourists

d. predators switch from small to large groups of animals of the same species to increase the probability of a successful kill

8.17a The functional response of a predator to prey availability is illustrated in Fig. 8.9. Which specific type of response does this illustrate?

a. Type I

b. Type II

c. Type III

d. Type IV

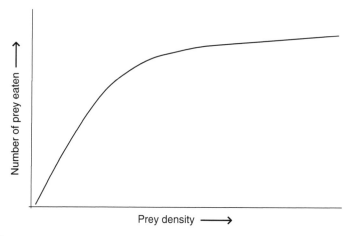

Fig. 8.9.

8.18a **According to optimal foraging theory, when should a predator add a new prey item to its diet?**

   a. When the new prey item increases the predator's rate of energy intake compared with that from its previous diet

   b. When the new prey item contributes to a more balance diet

   c. When the new prey item becomes scarcer

   d. When biodiversity in the area is high

8.19a **Robert MacArthur suggested that community stability is related to the number of trophic pathways in an ecosystem. Trophic pathways are**

   a. potential routes of nutrient cycling within an ecosystem

   b. potential interaction between members of the same trophic level in an ecosystem

   c. potential routes of energy flow from the producer to the highest consumer trophic levels in an ecosystem

   d. potential routes of energy flow from the producer to the first consumer trophic level in an ecosystem

8.20a **Sheep liver fluke disease is caused by**

   a. *Toxoplasma gondii*

   b. *Schistosoma mansoni*

   c. *Giardia lamblia*

   d. *Fasciola hepatica*

9 Ecological Genetics and Evolution

This chapter contains questions about the relationship between the genetics and ecology of organisms, their adaptations to the environment, and their evolution.

Foundation

9.1f   The ultimate source of all diversity in living organisms is

    a. the recombination of genes during sexual reproduction

    b. genetic drift

    c. natural selection

    d. mutation

9.2f   Kangaroos in Australian grasslands occupy similar ecological niches to those occupied by antelopes in African savannahs. This is an example of

    a. stabilising selection

    b. convergent evolution

    c. disruptive selection

    d. divergent evolution

9.3f The range of species K in Fig 9.1a was split in two by the emergence of a high mountain range forming populations K_1 and K_2 (Fig. 9.1b). Thereafter population K_1 evolved into species L and population K_2 evolved into species M (Fig. 9.1c), because populations K_1 and K_2 could no longer interbreed due to genetic changes that occurred within them. Which of the following terms is not used to describe this type of speciation?

   a. Vicariant speciation

   b. Geographical speciation

   c. Sympatric speciation

   d. Allopatric speciation

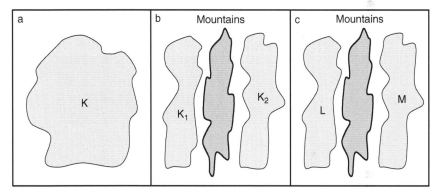

Fig. 9.1.

9.4f Some harmless (palatable) species have evolved so that they exhibit the same warning signals as harmful (toxic) species to protect themselves from predators. For example, harmless hornets have a yellow and black striped abdomen similar to that of wasps. This phenomenon is known as

   a. Müllerian mimicry

   b. Batesian mimicry

   c. Browerian mimicry

   d. Darwinian mimicry

9.5f **Charles Darwin's theory of evolution by natural selection is similar to a contemporaneous theory developed by**

 a. Alfred Russel Wallace after studying organisms in Southeast Asia

 b. Richard Owen after studying dinosaur skeletons

 c. Erasmus Darwin after studying fossils

 d. Stamford Raffles after exploring the jungles of Singapore

9.6f **Drought-resistant crop plants have been developed by**

 a. natural selection

 b. disruptive selection

 c. artificial selection

 d. inbreeding depression

9.7f **Fig. 9.2 shows a sample of *Partula* snail shells. The variation shown by these shells is the raw material of**

 a. random assortment

 b. natural selection

 c. disruptive selection

 d. systematics

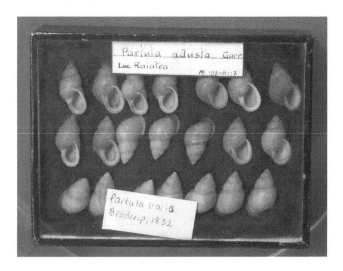

Fig. 9.2.

9.8f The diversity of forms found within the Insecta is illustrated by Fig. 9.3. This is an example of

   a. convergent evolution

   b. coevolution

   c. adaptive radiation

   d. parallel evolution

Fig. 9.3.

9.9f An interbreeding local group of a species whose members share a gene pool is known as a

   a. deme

   b. cline

   c. clone

   d. meme

9.10f **The pool of deleterious genes in a population is referred to as its**

a. genome

b. genetic load

c. genetic pollution

d. gene bank

9.11f **The genetic structure of a population is determined by**

a. selection and recombination

b. recombination and mutation

c. mutation and selection

d. selection, recombination and mutation

9.12f **If there is a small amount of gene flow among populations these populations will**

a. evolve almost independently

b. evolve together

c. become extinct

d. exhibit genetic drift

9.13f **Organisms that are well adapted to their environment tend to be more likely to survive, and produce more offspring, than those that are less well adapted to their environment. This is a description of the process known as**

a. coevolution

b. natural selection

c. adaptation

d. evolution

9.14f **Merging of the morphological characteristics of two populations when they come into contact is evidence that they are not reproductively isolated and should be treated as a single species. This process is called**

a. inbreeding

b. interbreeding

c. intergrading

d. speciation

9.15f Alteration of the gene pool by genetic drift, gene migration and natural selection is known as

a. microevolution

b. neutral selection

c. macroevolution

d. genetic equilibrium

9.16f In which decade did Charles Darwin publish *On the Origin of Species*?

a. 1850s

b. 1860s

c. 1870s

d. 1880s

9.17f A phenotypic trait that has evolved to assist an organism in dealing with some aspect of its environment is

a. an adjustment

b. a modification

c. an alteration

d. an adaptation

9.18f The capacity of an organism to pass its genes to the next generation is referred to as its

a. fecundity

b. fitness

c. vitality

d. vigour

9.19f The science of genomics is concerned with

a. evolution

b. gene mapping

c. gene editing

d. all of the above

9.20f Ecological genetics is the study of genetics in

a. laboratory populations

b. natural field populations

c. animal populations

d. plant populations

Intermediate

9.1i The last three individuals in a population of birds that possessed the allele for a particular character were killed during a severe storm. The process by which this allele was removed from the population is known as

a. genetic extinction

b. genetic drift

c. genetic flow

d. genomic drift

9.2i Individual peppered moths (*Biston betularia*) appear light grey in clean environments and black where environments have been affected by soot from air pollution deposited on tree bark and buildings. This is an example of

a. industrial melanism

b. industrial albinism

c. convergent evolution

d. stabilising selection

9.3i When body length was measured in a population of small songbirds it was found to exhibit a normal (bell-shaped) distribution. After a particularly cold winter the mean body length in the population was found to have shifted to the right. This is an example of

a. disruptive selection

b. directional selection

c. artificial selection

d. stabilising selection

9.4i **Species A and species B have been formed from the same ancestral species within the same geographical range. These two species are capable of interbreeding but do not due to ecological barriers. They have arisen as a result of**

a. allopatric speciation

b. parapatric speciation

c. sympatric speciation

d. convergent speciation

9.5i **In captive-breeding programmes, a decline in the viability or fertility of individuals in a population may result from mating between close relatives. This is known as**

a. inbreeding reversion

b. inbreeding suppression

c. inbreeding depression

d. inbreeding regression

9.6i **Add the missing term to the sentence below:**

The .... population size is defined as the size of an ideal population that would lose genetic variation by genetic drift at the same rate.

a. effective

b. affective

c. effectual

d. selective

9.7i **A metallophyte is a type of plant that**

a. is heavy metal tolerant

b. is susceptible to poisoning by low levels of metals in the soil

c. requires high levels of several metals to maintain good health

d. has a very high requirement for one particular metal

9.8i The study of the processes that control the geographical distributions of lineages by constructing the genealogies of populations and genes is called

    a. genealogy

    b. phylogenealogy

    c. phylogeny

    d. phylogeography

9.9i What percentage of a male baboon's genes has he inherited from his paternal grandfather?

    a. 100%

    b. 50%

    c. 25%

    d. 12.5%

9.10i The ability of animals to recognise genetic relatives is known as

    a. gene recognition

    b. family recognition

    c. kin recognition

    d. relative recognition

9.11i The uncontrolled flow of genes into wild populations from domestic, feral, invasive or genetically-engineered species may threaten the existence of some species and is known as

    a. genetic mixing

    b. genetic pollution

    c. genetic swamping

    d. all of the above

9.12i A fragment of DNA associated with a certain location within the genome and used to identify a particular DNA sequence in a pool of unknown DNA is called a

a. locus

b. molecular marker

c. meme

d. cline

9.13i The phenomenon whereby some individual animals are able to attract more mates than others by being more attractive to the opposite sex is known as

a. stabilising selection

b. disruptive selection

c. kin selection

d. sexual selection

9.14i The ability of one genotype to produce more than one phenotype in response to different environments is known as

a. phenotypic plasticity

b. phenotypic elasticity

c. genotypic plasticity

d. environmental plasticity

9.15i The lion (*Panthera leo*) population of Ngorongoro crater in Tanzania is isolated from those in the surrounding areas and the strong coalitions formed by males may have prevented immigrants from entering the crater. This may have resulted in an increase in

a. natality

b. outbreeding

c. mortality

d. inbreeding

9.16i **Which of the following are considered to be the pioneers of population genetics theory?**

a. Haldane, Mayr and Ford

b. Fisher, Haldane and Wright

c. Wright, Ford and Fisher

d. Hardy, Haldane and Ford

9.17i **Who was the first scientist to study the evolutionary mechanism behind the adaptation of moths to changes in the environment caused by atmospheric pollution by conducting field experiments?**

a. Bernard Kettlewell

b. Philip Sheppard

c. Arthur Cain

d. Laurence Cook

9.18i **In the wild in North America, pizzly bears have been produced from matings between**

a. black bears (*Ursus americanus*) and brown bears (*U. arctos*)

b. polar bears (*Ursus maritimus*) and brown bears (*U. arctos*)

c. black bears (*Ursus americanus*) and polar bears (*U. maritimus*)

d. polar bears (*Ursus maritimus*) and sloth bears (*Melursus ursinus*)

9.19i **Kin selection is a form of natural selection that favours genes that promote**

a. a high reproductive rate

b. altruistic acts that benefit relatives

c. a high rate of offspring survival

d. adaptation to a wide range of habitats

9.20i Which of the following techniques is used to make large numbers of copies of DNA for analysis?

    a. Chromatography

    b. Differential centrifugation

    c. Electrophoresis

    d. Polymerase chain reaction

Advanced

9.1a Which of the following conducted the detailed work on resource utilisation and adaptive radiation in Galapagos finches (*Geospiza* spp.)

    a. Charles Darwin

    b. Robert FitzRoy

    c. Edward O. Wilson

    d. David Lack

9.2a Outbreeding depression is the result of the mating of

    a. distant relatives

    b. closely related species

    c. siblings

    d. hybrids

9.3a If a population of reindeer (*Rangifer tarandus*) consists of 109 males and 141 females the effective size of this population is

    a. 250

    b. >250

    c. <250

    d. 141

9.4a During its evolution, the population size of a species may become greatly reduced due to an adverse environmental factor, such as a change in climate or the emergence of a new disease. This population may later expand in size but is said to have experienced a genetic bottleneck. This may result in

a. a reduction in biodiversity

b. an increase in genetic diversity

c. an increase in mutation rate

d. a reduction in genetic diversity

9.5a Many species exhibit gradual changes in their morphology from one end of their range to the other. The coconut lory (*Trichoglossus haematodus*) is a parrot that occurs in New Guinea. From the north to the south of its range the blue of its head becomes brighter, while its yellow collar becomes greener and less distinct from west to east. These smooth character gradients are called

a. clones

b. clines

c. clades

d. memes

9.6a The genetic relationships within five types of organisms based on the sequential order in which branches arise from a phylogenetic tree are illustrated in Fig. 9.4. It takes the form of a series of dichotomous branches. Each branching point is defined by the new homologous characters that are unique to the species on each branch. The diagram is most accurately called

a. an evolutionary tree

b. a genealogical map

c. a cladogram

d. an ancestral tree

9.7a. E.B. Ford was a leading British ecological geneticist who studied natural selection in

a. birds

b. beetles

c. butterflies and moths

d. ants

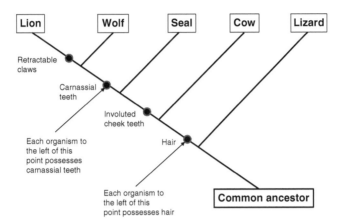

Fig. 9.4.

9.8a The distribution of a species of parrot that has spread from location A to location B, around a large mountain, over many thousands of years is illustrated in Fig. 9.5. Now, although they are close enough to encounter each other in the wild, male parrots taken from location A are unable to breed with females taken from location B. This is because, as the species' range extended away from location A anti-clockwise around the mountain, the individuals at the extremes of the range reached a point where they no longer interbred because the distance between them was too great and, as a result, they began to diverge genetically. This species is known as a

a. hoop species

b. band species

c. halo species

d. ring species

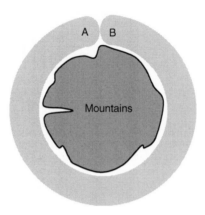

Fig. 9.5.

9.9a The small land snail *Cepaea nemoralis* occurs in a wide range of habitats. In each habitat the shell of the snail has a colour similar to that of its background. For example, on sandy soils the shell has a light colour. This phenomenon has most likely come about due to

  a. natural selection because birds are more easily able to find snails whose shell colour contrasts with the background habitat than snails that are camouflaged

  b. the absorption of different chemicals from the soil in each habitat which affects the colour of the shell as it grows

  c. the different foods consumed by the snails in the various habitats

  d. the effects of climate on shell development in the different habitats

9.10a The diagram below (Fig. 9.6) illustrates

  a. genetic drift

  b. the Hardy-Weinberg equilibrium

  c. natural selection

  d. inbreeding depression

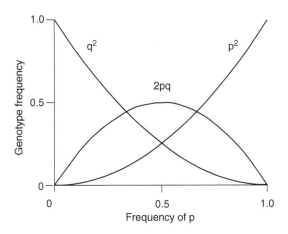

Fig. 9.6.

9.11a **Where the ranges of the carrion crow (*Corvus corone*) and the hooded or hoodie crow (*C. cornix*) meet in Europe and northern Asia there exists an area in which interbreeding occurs between the two species. The carrion crow is black and the hooded crow in black and grey. When they interbreed the resultant offspring exhibit a very wide range of variation in the distribution of black and grey feathers. This area is known as**

    a. an isolation zone

    b. a variation zone

    c. an interbreeding zone

    d. a hybrid zone

9.12a **The Isthmus of Panama is a narrow section of land that connects North and South America. A high proportion of the families of mammals present in Mexico have failed to disperse southwards across the isthmus and a high proportion of mammalian families in Guyana (in South America) have failed to disperse northwards across the isthmus. This is probably due to the limited supply of habitats in this area. This dispersal route is most accurately described as**

    a. a corridor

    b. a filter route

c. a sweepstake route

d. an immigration route

9.13a When a small population that branches off from a larger population is not typical of the parent group the smaller population eventually comes to have a different genetic makeup from the parent population. This phenomenon is known as

a. instantaneous speciation

b. ecological speciation

c. the founder principle

d. character displacement

9.14a Which of the following actions would be least likely to increase an organism's genetic fitness?

a. Leaving a large number of offspring

b. Assisting a sibling in rearing her young

c. Assisting a cousin in rearing her young

d. Assisting an unrelated member of her social group in rearing her young

9.15a The genotypes of individual plants in four populations with respect to a single gene (D/d) are shown in Fig. 9.7. In which of the four populations A, B, C and D has the recessive allele (d) become fixed?

a. A

b. B

c. C

d. D

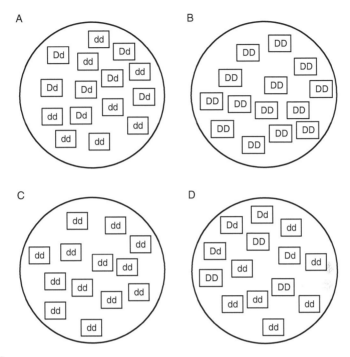

Fig. 9.7.

9.16a In a population that conforms to the Hardy-Weinberg equilibrium, if the frequency of the recessive allele (a) is 0.3 in the current generation, what will be the frequency of the heterozygotes (Aa) in the next generation?

a. 0.49

b. 0.42

c. 0.09

d. 0.21

9.17a In his book *Animal Species and their Evolution* Prof. Arthur Cain wrote '...they provide the finest demonstration of the progressive change of whole populations with distance, resulting in the attainment of specific status – that is, of geographical speciation' (Cain, 1971). What was he writing about?

a. Polymorphic species

b. Ring species

157

c. Clines

d. Species groups

9.18a Which of the following is not an assumption made by the Hardy-Weinberg equilibrium?

a. There is no immigration into the population

b. No mutation occurs in the population

c. Mating within the population is non-random

d. No selection occurs within the population

9.19a In a population that conforms to the Hardy-Weinberg equilibrium, if the frequency of the double recessive genotype (ff) is 0.28, what is the frequency of the double dominant genotype (FF)?

a. 0.53

b. 0.47

c. 0.72

d. 0.22

9.20a Once an evolutionarily stable strategy (ESS) is fixed in a population alternative strategies are prevented from successfully invading by

a. natural selection

b. mutation

c. random assortment

d. immigration

10 Ecological Methods and Statistics

Foundation

This chapter contains questions about the methods used by ecologists to sample and measure the biological components of ecosystems, and the statistical analysis of ecological data.

10.1f   A Baermann funnel is designed to

   a. sample flying insects

   b. extract organisms that live in soil water films

   c. collect organisms that occupy soil air spaces

   d. filter small freshwater invertebrates from samples of river water

10.2f   A spring scale is most likely to be used to

   a. weigh a small bird or other animal

   b. measure the length of a bird's wing

   c. trap a small animal

   d. measure plant growth in spring

10.3f   A place where migratory birds are regularly captured and ringed so that their future movements may be monitored is called

   a. an ornithological station

   b. a bird station

   c. a bird observatory

   d. a migration station

10.4f **An autonomous submersible vehicle would be most useful for studying animals living in**

    a. terrestrial forest environments

    b. deep ocean environments

    c. shallow riverine environments

    d. aerial environments

10.5f **A major advantage of using a Sherman trap rather than another type of small mammal trap is that it**

    a. folds flat, making it easier to store and transport

    b. is made of plastic

    c. kills animals humanely

    d. does not need to be inspected regularly once deployed

10.6f **The device in Fig. 10.1 is located in a cloud forest and used to collect**

Fig. 10.1.

a. rainfall

b. soil invertebrates

c. flying insects

d. falling tree leaves

10.7f A grapnel is a three-pronged metal hook which is generally used to sample

a. terrestrial plants

b. aquatic plants

c. invertebrates

d. soil

10.8f A metal case designed for temporarily storing and protecting plants collected in the field is called

a. a herbarium

b. a vasculum

c. a botanical case

d. an arboretum

10.9f Insects living in the aerial parts of bushes and trees may be sampled by striking the branches with a wooden stick and catching the fallen animals in a

a. moth trap

b. mist net

c. beating tray

d. sweep net

10.10f Ecological studies that use members of the general public to collect data are generally referred to as

a. citizen science projects

b. people science projects

c. popular science projects

d. public science projects

10.11f A device that consists of a series of metal pins suspended from a linear frame and used to sample vegetation growing on the surface of the ground is called a

a. sampling frame

b. linear frame

c. point frame

d. pin frame

10.12f A harp trap is used for the live capture of

a. birds

b. small mammals

c. fishes

d. bats

10.13f The technique known as 'knock-down fogging' is most likely to be used to collect samples of

a. plants

b. insects

c. small mammals

d. seeds

10.14f A seine net may be used to catch

a. butterflies

b. freshwater invertebrates

c. fishes

d. bats

10.15f A Longworth trap is used to

a. catch small mammals alive

b. catch and kill small mammals

c. catch and kill small birds

d. catch small birds alive

10.16f An ecologist who wishes to catch small songbirds alive so that they may be fitted with identification rings would probably use a

a. Larsen trap

b. mist net

c. gin trap

d. spring trap

10.17f A pooter is a small portable device used for collecting

a. small terrestrial insects

b. small freshwater invertebrates

c. small mammals

d. small plants

10.18f A pitfall trap is used to sample

a. zooplankton

b. terrestrial invertebrates

c. birds

d. aerial insects

10.19f An ecologist studying primary production in a woodland measured the diameter of acorn seeds to the nearest millimetre and then calculated the mean, mode, median, variance and standard deviation of the values obtained. Which of the following pairs of terms are measures of dispersion?

a. Mean and mode

b. Mean and standard deviation

c. Median and standard deviation

d. Variance and standard deviation

10.20f A field biologist studying a population of small songbirds made recordings of four of their morphological characteristics: beak length, body mass, plumage colour and wingspan. Which of these is not a continuous variable?

a. Beak length

b. Body mass

c. Plumage colour

d. Wingspan

Intermediate

10.1i **Which of the following is used to obtain invertebrates from a soil sample?**

a. A Petersen funnel

b. A Tullgren funnel

c. A Longworth funnel

d. A Larsen funnel

10.2i **The quantity of energy in a sample of biological material may be measured by using the process known as**

a. spectroscopy

b. bomb calorimetry

c. differential centrifugation

d. chromatography

10.3i **A sample of invertebrates collected from the bottom of a river by disturbing the gravel with one foot while holding a net downstream to catch any animals that are dislodged is called a**

a. kick sample

b. foot sample

c. net sample

d. disturbance sample

10.4i **Which of the following statements about binoculars described as 8 x 42 is false?**

a. They have a magnification of x 8

b. The largest lenses have a diameter of 42mm

c. They have the same magnification as a field telescope described as 8 x 70

d. They have a narrower field of view than binoculars described as 10 x 25

10.5i **Which of the following specifications for binoculars is likely to be least useful for general ecological fieldwork such as counting birds or observing large mammals?**

a. 8 x 35

b. 6.5 x 35

c. 8 x 42

d. 12 x 56

10.6i **Which of the following sampling methods does not involve the use of a transect?**

a. The population density of deer in a park is estimated by walking in a straight line for 2km and counting the deer observed within 200m either side of the line

b. The distribution of flowering plants was recorded by assessing percentage cover of each species within quadrats placed at 15m intervals in a straight line across a salt marsh

c. The distribution and abundance of various species of snails were determined in an area of mixed woodland by examining 40 randomly located quadrats

d. The density of zebra in an area of savannah was estimated by counting animals observed 1km either side of the flight path of a light aircraft flown due south for 25km

10.7i **A population of badgers (*Meles meles*) consists of 175 males and 197 females. Which of the following statistical tests would you use to determine whether or not these values differ from a 1:1 ratio of males:females?**

a. Independent t-test

b. Correlation coefficient

c. Chi-squared test

d. Dependent t-test

10.8i **Which of the following statements about quadrats is false**

a. They may be used to sample populations of plants and sessile animals

b. They may be of any size

c. They must be square

d. They may be used to calculate plant density

10.9i **An ecologist weighed 150 field voles (*Microtus agrestis*) and found that the weights of the animals exhibited a normal distribution. In such a distribution which of the following statements is true?**

a. Only the mean is located in the centre of the distribution

b. The mean, median and mode are all located in the centre of the distribution

c. The mean and mode are located in the centre of the distribution

d. The median and the mean are located in the centre of the distribution

10.10i An ecologist sampled a population of a plant species in a grassland habitat using a one metre square quadrat thrown randomly. Table 10.1 shows the number of quadrats that contained between 0 and 9 individuals of this species.

Table 10.1

Number of plants per quadrat	Number of quadrats
0	69
1	0
2	1
3	2
4	0
5	3
6	12
7	9
8	15
9	17

The distribution of this species is best described as

a. regular

b. random

c. normal

d. clumped

10.11i Which of the following is most useful for determining the presence of a rare species in a river system from a water sample?

a. eDNA

b. mDNA

c. tRNA

d. rRNA

10.12i An ecologist counted the number of daisy plants in eight 1m² quadrats sampled in an area of 55m². Assuming that the distribution of the plants in the areas sampled was representative of that throughout the population under study, calculate an estimate of the population size using the data in Table 10.2.

Table 10.2

Quadrat	Number of daisy plants
1	22
2	17
3	8
4	35
5	0
6	28
7	13
8	11

a. 857

b. 950

c. 883

d. 921

10.13i An ecologist took a soil core (cylinder) from a habitat and examined the plant remains that occurred at different levels within it. The top layer contained tussock-forming sedges. Underneath this she found sedge peat, and at the base of the core the remains of reeds were found. This provides evidence of

   a. the zonation of plants in a lake

   b. a succession that began with tussock-forming sedges

   c. the zonation of plants in a marsh

   d. a succession that began with reeds

10.14i Estimates of the population of giraffes (*Giraffa camelopardalis*) in an African national park, where the habitat is savannah, are shown in Table 10.3. Which of the following statements is supported by the evidence?

Table 10.3

Year	Survey method	Population estimate
1985	Strip transect samples by aircraft	2186
1990	Total count from aircraft	2301
1996	Strip transect samples using vehicles confined to roads	1947
2000	Total count from aircraft	2521

   a. The population increased between 1985 and 2000

   b. The population decreased between 1990 and 1996

   c. The population increased between 1996 and 2000

   d. The population increased between 1990 and 2000

10.15i The 'unopened holes' method may be used to estimate the size of a population of

   a. small desert rodents

   b. forest songbirds

c. marine crabs

d. tropical butterflies

10.16i An experiment conducted using a computer to examine the effect of manipulating equations that are believed to describe real-world ecological processes is called

a. a paradigm

b. a replication

c. a duplication

d. a simulation

10.17i Many early mathematical descriptions of population processes were developed by scientists whose background was in

a. ecology

b. the physical sciences

c. zoology

d. botany

10.18i The logistic equation may be used to study one type of population growth and is an example of

a. a mathematical estimation

b. a mathematical theorem

c. a mathematical model

d. a mathematical hypothesis

10.19i A bathyscope is

a. a field microscope

b. used for night viewing

c. used for underwater viewing

d. used for viewing microscopic organisms

10.20i The following is a list of the numbers of small snails found in 15 quadrats thrown randomly in an area of woodland ranked from the lowest to the highest: 0, 1, 1, 1, 1, 1, 2, 2, 2, 2, 3, 3, 3, 4, 36.

Which of the following is the most appropriate measure of central tendency to represent the 'middle value' for these data?

a. The median

b. The mean

c. The mode

d. The variance

Advanced

10.1a **In a study conducted over 75 days Prof. Todd found a correlation coefficient of zero between the maximum daily temperature (variable A) and the number of moths captured in a moth trap (variable B) indicating that**

a. as A increased B increased

b. as A increased B decreased

c. there was no relationship between changes in A and changes in B

d. none of the above

10.2a **Dr Harrison suggested to one of her students that a type I error had occurred in his data analysis. This meant that he had**

a. detected an effect that was not present

b. failed to detect an effect that was present

c. made a mathematical error in his calculations

d. made an error equivalent to the standard error

10.3a **While conducting a large mammal survey from a road vehicle in India an ecologist located a tiger (*Panthera tigris*). To estimate the GPS position of the tiger at the time of the sighting he would need to determine**

a. the GPS location of the vehicle and the distance between the vehicle and the tiger

b. the GPS location of the vehicle and the compass bearing (in degrees) from the vehicle to the tiger

c. the GPS location of the vehicle, the distance between the vehicle and the tiger, and the compass bearing (in degrees) from the vehicle to the tiger

d. the distance between the vehicle and the tiger, and the compass bearing (in degrees) from the vehicle to the tiger

10.4a An ecologist studied the population dynamics of a small desert antelope by counting the number of individuals killed each month by motor vehicles on a major road and using this as a proxy for population size. The results for a single year are presented in Table 10.4. Which of the following statements is false?

a. The method used takes into account monthly changes in traffic frequency

b. Assuming the methodology is valid the data could only be used to determine relative changes in antelope numbers not actual changes in the population size

c. If the methodology used is valid, the population in May appears to have been approximately one third of that in April

d. A doubling of traffic volume from one month to the next would mean the number killed would double if the population size stayed the same

Table 10.4

Month	Jan	Feb	Mar	Apr	May	June	Jul	Aug	Sep	Oct	Nov	Dec
Deaths	33	27	21	29	9	12	16	11	14	26	32	29

10.5a An ecologist measured the invertebrate biodiversity of pond A using Menhinick's index and the invertebrate biodiversity of pond B using Simpson's index. The values obtained were 0.67 for pond A and 0.49 for pond B. Which of the following statements is true?

a. The biodiversity of invertebrates is higher in pond A than in pond B

b. The biodiversity of invertebrates is higher in pond B than in pond A

   c. Simpson's index should not be used to measure invertebrate biodiversity

   d. It is impossible to say which pond has the higher invertebrate biodiversity from the information available

10.6a An ecologist wants to compare the diameter of limpets (Gastropoda) found on a beach which has been polluted by oil with those found on one nearby that is unpolluted because he suspects that the oil has reduced their growth rate. He has 250 measurements of limpets from each beach. Which of the following statistical tests should he use?

   a. Regression analysis

   b. Dependent t-test

   c. 2x2 contingency table

   d. Independent t-test

10.7a In a field study of the effect of an environmental variable on plant growth in a tropical forest an ecologist used regression analysis to examine the data. This essentially calculates

   a. the skewness of a distribution

   b. the position of a line of best fit through a series of points on a graph

   c. the dispersion of values around the mean

   d. the extent to which the difference between the means of two distributions is statistically significant

10.8a A mathematical model of an ecological process whose outcome depends upon the effect of random influences may be described as

   a. deterministic

   b. linear

   c. exponential

   d. stochastic

10.9a The numbers of individuals in each of four species (K, L, M and N) in four animal communities (A, B, C and D) are shown in Table 10.5. Which community has the lowest species evenness?

a. A

b. B

c. C

d. D

Table 10.5

Community	Species			
	K	L	M	N
A	65	12	15	43
B	17	21	15	24
C	4	8	57	2
D	16	72	51	32

10.10a The size of a population of plants may be estimated by first calculating their density from the numbers of individuals recorded within randomly thrown quadrats. If only a small number of quadrats is thrown the population estimate is likely to be inaccurate because you could sample areas with only high plant densities or only low plant densities. A larger number of quadrats is likely to provide a more accurate estimate because the extremes of density would be averaged out, but if a very large number of quadrats is thrown there will not be a concomitant improvement in accuracy and extra effort could be expended for nothing. Fig. 10.2 illustrates the relationship between sample size (number of quadrats thrown) and population estimate for a plant population. How many quadrats would you need to throw to obtain an accurate estimate of the population size without expending unnecessary effort?

a. 4

b. 5

c. 12

d. 18

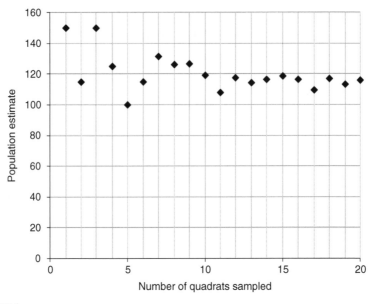

Fig. 10.2.

10.11a Nearest neighbour analysis is a type of

    a. plotless sampling

    b. random sampling

    c. regular sampling

    d. stratified sampling

10.12a When the distribution of a population of plants is examined using random quadrats the relationship between the mean number of plants per quadrat and the variance can be used to measure the degree of clumping by calculating an index of dispersion (I) where:

$$I = \frac{variance}{mean}$$

The value of *I* can be interpreted as indicated in Table 10.6.

The distribution of four plant populations are shown in Fig. 10.3. The four areas (A-D) represent habitats of identical size. Which population would you expect to have the highest value for *I*?

Table 10.6

Type of distribution	Relationship between mean and variance	Value of *I*
Random	Variance = mean	1
Uniform/regular	Variance < mean	< 1
Clumped	Variance > mean	> 1

a. A

b. B

c. C

d. D

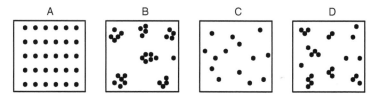

Fig. 10.3.

10.13a An ecologist ran a Monte Carlo simulation of microbial population growth on a computer. Each time the simulation was run we would expect the outcome to be

a. larger than the last time it was run

b. smaller than the last time it was run

c. different from the last time it was run

d. the same as the last time it was run

10.14a Herring gulls (*Larus argentatus*) and lesser black-backed gulls (*L. fuscus*) nest on the ground in multi-species colonies in sand dune ecosystems. Some individuals establish territories on the dunes themselves and others in the sand dune slacks between them. Which test would you use to determine whether or not these gull species prefer different habitats?

a. A 2x2 contingency table

b. Regression analysis

c. A dependent t-test

d. A cluster analysis

10.15a An ecologist put samples of leaf litter from the same site in three mesh bags – each of which had a different mesh size (small, medium and large) – and then placed them in the soil. The purpose of this experiment was to determine the effect of which factor on the rate of decomposition of the litter?

a. Water content

b. Microclimate

c. Bacteria and fungi

d. Different types of decomposers

10.16a Which of the following would you use to perform faecal worm egg counts under a microscope?

a. A McMaster counting slide

b. A McAvoy counting slide

c. A McKinleigh counting chamber

d. A McDougall counting chamber

10.17a An area of land (A) occupied by a population of deer is shown in Fig. 10.4. Areas A, B and C show the locations of three transects used to sample the population. The population size can be estimated by sampling a number of transects using the formula:

$$Population\ estimate(N) = \frac{total\ study\ area}{transect\ area} \times \frac{mean\ number}{of\ deer\ per\ transect}$$

If the total area is 160ha (1,600,000 m²), $L = 700$m, and $w = 75$m, the population estimate is

a. 86

b. 91

c. 105

d. 117

Total area (A)

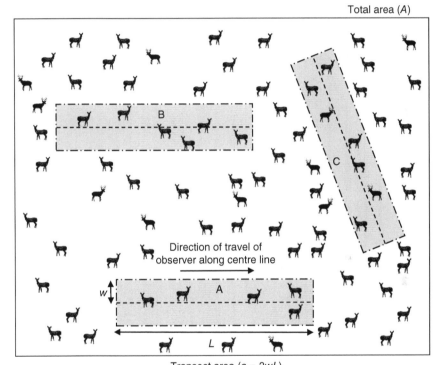

Transect area ($a = 2wL$)

Fig. 10.4.

10.18a Field biologists sometimes use devices that play recordings of bat sounds to attract bats during a census. These devices are called

a. acoustic snares

b. acoustic decoys

c. acoustic baits

d. acoustic lures

10.19a A kite diagram would be most useful in illustrating

    a. the relationship between shoulder height and body mass in giraffes

    b. the distribution of tail lengths in a population of songbirds

    c. the abundance of different plant species along a transect of a salt marsh

    d. the diversity of species in a pond

10.20a The dendrogram shown in Fig. 10.5 shows the similarities between nine different soil types (A–I) based on an analysis of their characteristics. It has been produced as the result of

    a. a regression analysis

    b. a cluster analysis

    c. a Monte Carlo simulation

    d. an analysis of variance

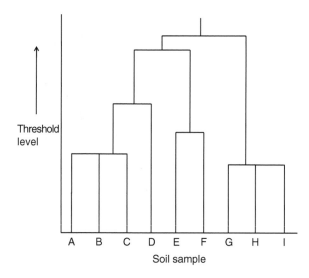

Fig. 10.5.

11 Answers

A multiple choice question has a stem (the 'question'), a key (the 'answer') and a number of distracters (wrong answers intended to distract the student from the key). This part of the book contains the key to each question along with a brief explanation of why this is correct and, in some cases, what the distracters mean.

Chapter 1 The History and Foundations of Ecology

1.1f	B	The term was first used by the German zoologist Ernst Haeckel in his book *Generelle morphologie der organismen*.
1.2f	C	First used in 1866 by Ernst Haeckel.
1.3f	D	*Oikos* is the Greek word for 'home'.
1.4f	A	This is a widely used basic definition but many others exist.
1.5f	B	Autecology is the study of the ecology of a single species. Synecology is the study of the ecology of whole animal or plant communities. The other distracters are fictitious.
1.6f	B	The English botanist Arthur Tansley first used the term ecosystem in a paper in 1935. Eugene Odum and Charles Elton were both famous ecologists.
1.7f	D	The ecosystem concept includes artificially created systems including aquariums and even space ships.
1.8f	A	An anthropogenic factor is one resulting from human activity, for example pollution, deforestation, over-fishing.

1.9f	A	An organism is the smallest unit relevant to an ecologist. A small isolated group of organisms that share a gene pool is a deme. A population is a group of individuals of the same species. Species that live together form a biological community.
1.10f	C	Rev. Gilbert White wrote *The Natural History and Antiquities of Selborne*
1.11f	B	*A Sand County Almanac* was Leopold's book which was published in 1949 and important in the development of environmental ethics. The other titles are fictitious.
1.12f	B	Most life on Earth occurs within the temperature range 0°C – 50°C.
1.13f	D	Biogeography is the study of the origin, distribution, adaptation and associations of plants and animals. The equivalent term for animals is zoogeography.
1.14f	B	An Eltonian pyramid is another name for an ecological pyramid (of numbers, energy or biomass) used in the study of food chains and energy flow in ecosystems.
1.15f	D	Alexander von Humboldt was a German naturalist and geographer.
1.16f	A	One of Aldo Leopold's areas of expertise was wildlife management.
1.17f	B	The term 'conservation' implies the sustainable use of resources, whereas 'preservation' suggests they are protected from exploitation.
1.18f	D	The flora of an area is its plant life (e.g. the flora of France). *A Flora of Kenya* would be a field identification guide to the plants of Kenya or possibly a list of plant species.
1.19f	C	The term sward is used to refer to an area of ground covered in short grass and other small plants.
1.20f	C	Ecologists and foresters refer to a stand of trees. The term is also used to refer to crops.
1.1i	A	This was an important book written about human population growth by Malthus.
1.2i	D	The troposphere is the lowest layer of the atmosphere and is where all weather occurs. Moving upwards, the stratosphere occurs next, then the mesosphere and finally the thermosphere.
1.3i	C	The epipelagic zone is the surface layer of the ocean. Beneath this is the mesopelagic zone, followed by the bathypelagic and the abyssopelagic is beneath this.
1.4i	C	The hydrosphere is made up of all of the water on Earth, wherever it occurs.
1.5i	A	A littoral zone is on or near the shore of a lake or sea.
1.6i	B	Limnology is the study of inland aquatic systems; the biology, chemistry and physical features of rivers, lakes, etc.

1.7i	D	The term 'microcosm' is used to describe artificial, simplified ecosystems; the term 'biocoenosis' was coined by Karl Möbius in 1877; 'biogeocoenosis' is a term used in Russia and Central Europe attributed to V. Sukachaev who coined it in 1947.
1.8i	A	The nekton is made up of actively swimming organisms in a body of water that move independently of water currents, e.g. fishes, cetaceans, jellyfishes, etc.
1.9i	C	The Gaia hypothesis was proposed by James Lovelock. The distracters are fictitious hypotheses named after other females from Greek mythology.
1.10i	A	Wytham Wood has been widely used as a study site by ecologists, especially from the University of Oxford.
1.11i	B	Silver Springs was used for early studies of energy, especially by H. T. Odum in the 1950s.
1.12i	C	These scientists have made detailed studies of, and written books about, these species.
1.13i	A	Animal ecologists focussed on population ecology and plants ecologists focussed on community ecology in the early days of ecology. Plant population dynamics were not well understood.
1.14i	D	Some plants live much longer than animals, they produce dormant seeds and some reproduce vegetatively (making it difficult to define individuals). These features of plants make their population dynamics difficult to study. The fact that most plants are stationary makes them relatively easy to find compared with animals, but that does not make them easy to study.
1.15i	A	Brazil is the only country listed that is not part of the Old World.
1.16i	D	Giant pandas predominantly eat bamboo; koalas eat *Eucalyptus* species; ospreys eat fishes; Asian elephants consume a great number and variety of plant species.
1.17i	D	Rachel Carson's book *Silent Spring* focussed on the effects of chemicals on food chains.
1.18i	B	In the mid-1900s, European ecologists were interested in the composition, structure and distribution of plant communities and American ecologists were interested in the development of plant communities by the process of succession.
1.19i	A	Elton wrote *Animal Ecology*. It was published in 1927 and was important in helping to establish ecology as a science.
1.20i	B	The original meaning of the term 'natural history' included the physical components of the environment as well as the biological components. Modern usage tends to be restricted to the biological components.
1.1a	A	Fossorial animals have a burrowing habit, e.g. aardvarks, many rodents.
1.2a	D	The Ethiopian realm only extends to Africa south of the northern boundary of the Sahara desert, southern Arabia and Madagascar.

1.3a	B	B is correct. The distracters are the same information rearranged.
1.4a	C	Gondwana (or Gondwanaland) split to form the land masses listed. Laurasia, Pangea and Tethys are the names of other land masses.
1.5a	A	Zoogeographical regions were first defined on the basis of their bird faunas by Philip Sclater, an English zoologist who was Secretary of the Zoological Society of London.
1.6a	D	The Neotropical realm includes all of South America where these taxa occur.
1.7a	B	African elephants occur in the Ethiopian realm and Asian elephants in the Oriental realm.
1.8a	A	Frederic Clements was an American plant ecologist who was a pioneer in the study of plant succession.
1.9a	B	Andrewartha and Birch wrote this book. The distracters give the names of other pairs of ecologists who have published important works.
1.10a	A	Wynne-Edwards was an English zoologist who wrote a book called *Animal Dispersion in Relation to Social Behaviour*. His ideas were based on a type of group selection.
1.11a	D	John Harper was a British biologist who founded the study of plant population dynamics.
1.12a	B	Historical ecology examines the effects of humans on wildlife and ecosystems over time.
1.13a	A	Scansorial animals are adapted for climbing; cursorial for running; volant for flying or gliding; arboreal for life in the trees.
1.14a	D	Charles Elton (1930) said this in his book *Animal Ecology and Evolution*. This was a reference to the fact that nature is always changing, e.g. ecological succession, fluctuations in population size, etc.
1.15a	B	The Bureau of Animal Population was founded by Charles Elton within the Zoology Department of the University of Oxford.
1.16a	C	On his voyage around the world Darwin did not visit India.
1.17a	D	The papers were by Darwin and Wallace. Neither was present. Wallace was in the field in Southeast Asia and Darwin was unwell and had just lost a child.
1.18a	A	Naturalised animals or plants are those that have become part of the native fauna or flora over time, e.g. Canada geese and rabbits in Britain.
1.19a	C	Sloane, an Irish doctor, donated his collection to the museum. Rothschild established a separate natural history museum at Tring. Owen was director of the British Museum (Natural History); Raffles was one of the founders of the Zoological Society of London.
1.20a	D	Phytosociologists proposed that different plant communities should be treated like separate taxa of organisms. Ethology is the study of animal behaviour in the field.

Chapter 2 Abiotic Factors and Environmental Monitoring

2.1f	D	Edaphic factors are soil factors such as pH, aeration, water content, etc.
2.2f	A	A Stevenson screen is a container that holds instruments such as thermometers in the shade and protected from the effects of wind at a standard height above the ground so that weather recordings can be standardised and compared between locations.
2.3f	C	Frost line is a fictitious term. The other three terms are all used synonymously.
2.4f	A	A soil augur is a metal device that is screwed into the soil and then withdrawn, thereby removing a soil core for analysis.
2.5f	C	Permanently frozen soil such as that found in the Arctic is called permafrost. Moraine and glacial drift are features of glacial landscapes. Permasnow is a fictitious term.
2.6f	B	A Secchi disk is a circular plate divided into black and white sections that is lowered into water to determine the depth to which light penetrates.
2.7f	D	The other options contain mosses (A), fungi (B) and soil (C) which contains bacteria, invertebrates and other organisms.
2.8f	B	Only samples 2 and 4 have pH values lower than 7.0 (i.e. are acidic).
2.9f	D	A choice chamber is a small plastic container which may be divided into two or more sections each with its own microhabitat (dark/light; humid/dry). Small invertebrates (e.g. woodlice) placed in the chamber select suitable habitats by moving to them.
2.10f	B	An anemometer consists of cups attached to a spindle that spins under the influence of air movements and the speed of rotation is converted into wind speed.
2.11f	C	A clinometer is a hand-held device used to measure angles from the horizontal.
2.12f	B	Lotic refers to running water environments such as streams and rivers.
2.13f	B	A white background makes it easier to find small invertebrates that are usually relatively dark in colour.
2.14f	C	Effectively this is the quantity of water vapour in the air as a percentage of the maximum it could hold.
2.15f	C	This term relates to the direction a slope is facing. This affects the amount of sunlight it receives, the temperature, persistence of snow, etc.
2.16f	B	The intertidal zone is the area between high and low tide. The benthic zone is at the bottom of a body of water; the pelagic zone is the body of water that is not near the bottom of a water body or near the shore; the strand line is a line of seaweed and debris marking a previous high water level on a shore.

2.17f	B	This is a 'forest' of kelp (a seaweed) anchored to the bottom of the sea.
2.18f	A	A potometer is a device consisting essentially of a glass tube filled with water that is attached to the cut stem of a plant so that the water is contiguous with that in the xylem of the plant. As water evaporates from the leaves of the plant, water is drawn along the glass tube and as its meniscus travels along a scale the rate of water uptake – and thus the rate of transpiration – can be calculated.
2.19f	D	An atmometer is similar in design to a potometer but is used to measure the rate of evaporation of water from a moist surface. It can be used to compare transpiration from leaves measured using a potometer with water loss from a physical system.
2.20f	B	BOD is a measure of the consumption of oxygen by microbes in water. It indirectly measures organic pollution because polluted water supports more microbes than unpolluted water and therefore uses more oxygen.
2.1i	D	When dry soil is heated strongly the humus content will burn off.
2.2i	B	A dimictic lake mixes vertically from the surface to the bottom twice each year (in autumn and spring).
2.3i	A	In a thermally stratified lake the thermocline is a layer of water where the temperature falls rapidly with depth. The position of this zone changes with the time of the year and it disappears in winter.
2.4i	B	The top layer of a thermally stratified lake is the epilimnion; the layer beneath this is the thermocline (or metalimnion); and the bottom layer is the hypolimnion.
2.5i	B	Ice has a lower density than liquid water so floats on the surface of bodies of water such as ponds, lakes and seas. This property is important because it allows animals to survive and feed under the ice surface.
2.6i	C	Temperature falls with distance from the Equator and with altitude and this is reflected in the location of the tree line.
2.7i	C	Pedogenesis is concerned with the processes by which soil is formed.
2.8i	B	The B horizon is the subsoil level.
2.9i	C	The lowest level at the bottom of a lake or other aquatic system is the profundal zone and it therefore receives the lowest levels of light.
2.10i	B	Both forest and marshland are present in the crater and the plants in these systems are large and contain a great deal of chlorophyll, compared, for example, with short grasses.
2.11i	A	A crown fire is one that spreads across the tops of trees in a forest, as opposed to a ground fire.
2.12i	B	The extent to which a soil can retain water is measure by a tensiometer. A potentiometer measures electromotive force; a hydrometer measures the relative density of liquids; and a hygrometer measures humidity.

2.13i	A	A vernier calliper is used for accurately measuring distance.
2.14i	A	The SI unit of illuminance is the lux. The other units are light related; the candela is the base unit of luminous intensity in the SI system; the lumen is the SI derived unit of luminous flux; the foot candle is an old, non-SI, unit of illumination.
2.15i	B	An emergence trap rests on the surface of standing water and captures insects as they emerge into the air.
2.16i	D	This type of net is pulled behind a boat and filters plankton from water which are drawn into a container on the right-hand side.
2.17i	A	A counting chamber is a device that is examined under a microscope and used to count organisms or other small objects within grid squares. Lines 3mm apart would only be suitable for counting relatively large plankton; pollen grains, bacteria and viruses are too small.
2.18i	B	Biodiversity consists of different types of organisms. These cannot be measured with a data logger. Air temperature, water flow and precipitation are all single variables that can be measured by thermometers, flow meters and rainfall gauges respectively and these data can be recorded digitally.
2.19i	A	Cold air is denser than warm air and sinks on clear nights, flowing down hills into valleys.
2.20i	A	The initials stand for: O = organic layer, A = surface horizon consisting of minerals and organic material, B = subsoil, C = substratum, R = bedrock.
2.1a	D	Fire acts as an agent of natural selection and selects in favour of fire-resistant species. It also destroys old plant growth, allowing new shoots to emerge, and it returns nutrient from leaf litter back to the soil.
2.2a	C	Coral can only form in relatively warm seas in the temperature range 23–29°C.
2.3a	D	An underwater microphone is called a hydrophone.
2.4a	D	An electrical current passes between an anode and cathode in the water and a net is used to catch the fish. Electric fishing may be undertaken by people in waders walking upstream or from a boat, so the boat is not essential.
2.5a	B	The graph shows changes in water discharge (K) with time (L). The storm creates an increase in discharge (M) above the normal (base flow) level (N).
2.6a	A	The specific heat of water is higher than that of land, not lower. This means that water absorbs and releases heat more slowly than land.
2.7a	D	An isohyet is a line joining points of equal precipitation (rainfall).
2.8a	A	An isotach is a line joining points of equal wind speed. An isobar joins points of equal pressure.

2.9a	A	Phenology is the science concerned with biological periodicity. An isophene links points of similar periodicity.
2.10a	B	The climate under the head of vegetation and above the soil surface is the microclimate.
2.11a	D	The process described is lateralisation. Podzolization is a complex process that involves dissolved organic matter and ions of aluminium and iron being moved from the upper to the lower parts of the soil; mineralisation is the decomposition of organic matter that results in the release of soluble inorganic chemicals that are available to be absorbed by plants.
2.12a	C	Savannah ecosystems experience high temperatures all year round and distinct wet and dry seasons.
2.13a	B	Snow reflects most light, followed by sand and then grass. Forest absorbs most light.
2.14a	B	BOD is measured by comparing oxygen use in dark and light bottles kept for 5 days at 20°C. This is a measure of the oxygen used by organisms in the water. The dark condition prevents oxygen being produced by photosynthetic organisms.
2.15a	D	The inverse-square law states that light intensity is inversely proportional to the square of the distance from the source.
2.16a	D	$10^9 = 1{,}000{,}000{,}000$ and $10^6 = 1{,}000{,}000$, so sample B is 1000 times more acidic than sample A.
2.17a	D	The stack of sieves should remove large particles first and allow smaller particles to fall through, so the sieve with the largest mesh size should appear at the top and that with the smallest at the bottom.
2.18a	B	A data buoy floats on the surface of water, especially in the ocean.
2.19a	A	The sievert is the SI unit of radiation dose. The other units are measures of radiation.
2.20a	C	A Burkard trap collects pollen and spores from the air. Such devices are used to provide daily pollen counts, often as part of a weather forecast.

Chapter 3 Taxonomy and Biodiversity

3.1f	A	Linnaeus was a Swedish physician at a time when those studying medicine also learned about many aspects of natural history. He practised medicine in Stockholm and later became Professor of Botany at Uppsala University in 1741.
3.2f	D	A lichen consists of an alga (or cyanobacterium) living in a symbiotic relationship with a fungus.

3.3f	B	These two species cannot be in different families because they are in the same genus and any genus can only be in one family.
3.4f	D	There are only about 5,500 mammal species, about 10,500 species of birds and reptiles, and over 7,500 species of amphibians.
3.5f	A	Endemic taxa are found in one area. For example, lemurs are endemic to Madagascar.
3.6f	D	The generic name of a species always begins with a capital letter but the specific name never does.
3.7f	A	This is the binomial name (consisting of two parts) or scientific name.
3.8f	C	In zoology, all family names end in 'idae', e.g. Felidae, Canidae.
3.9f	A	A dichotomous key leads the user through a series of questions to which there are only two answers until the species is identified.
3.10f	B	When taxa are revised in the light of new evidence about their evolutionary relationships they may be renamed, for example a species may be moved from one genus to another.
3.11f	A	This is the correct sequence. The other options have at least one taxon out of order.
3.12f	A	Continental drift is the movement of land masses on tectonic plates across the surface of the Earth.
3.13f	A	The application of distinctive names to organisms is nomenclature; systematics is the study of the kinds and diversity of organisms and their evolutionary relationships; classifcation is the grouping of organisms based on their similarities and/or evolutionary relationships; and taxonomy is the study of classification, its principles and rules.
3.14f	C	The term herpetology comes from the Greek root *herpeton* which means 'creeping thing'.
3.15f	A	The term icthyology is derived from the Greek for fish: *ikhthus*.
3.16f	D	Crabs are invertebrates and crustaceans and the only invertebrate zoologist in the group specialises in insects.
3.17f	D	Toads naturally occur in Canada but not in Tasmania, Madagascar or New Zealand.
3.18f	A	Plants are often preserved by drying and kept in a herbarium for identification purposes. An arboretum is a collection of trees.
3.19f	B	A ratite is a flightless bird. They are mostly large in size and have long legs, e.g. ostrich, rhea, emu, also kiwi (but not penguins).
3.20f	D	The specific name *niger* refers to the black coat colour. From the Latin root *nigr-*.
3.1i	C	The insects are part of this phylum and contain over a million species and the arachnids add over 100,000 additional species.

3.2i	B	The vernacular name is the common name so is quite different from the scientific name. A species may have many vernacular names.
3.3i	B	The rules of nomenclature require that the first subspecies has the specific name repeated, e.g. *Loxodonta africana africana*.
3.4i	D	A genus may contain a single species if it has no close relatives. For example the cheetah is the only extant member of the genus *Acinonyx*.
3.5i	A	There are over 1 million insect species, more than 33,000 fish species, about 10,500 species of birds, and only about 5,500 mammal species.
3.6i	B	The genus may be abbreviated to the first letter of the name if it has previously been mentioned.
3.7i	D	Occasionally new species are discovered in the wild. DNA evidence may be used to split previously known species into two or more species or to elevate a subspecies to the status of a species.
3.8i	B	Soil organisms >2mm are macrofauna, e.g. earthworms, ants, beetles, spiders. Mesofauna are 0.1 – 2mm. Megafauna are large species such as moles, snakes, gophers. Microfauna are microscopic species including protozoans and small nematodes.
3.9i	C	*Felis* spp. means two or more species of the genus *Felis*.
3.10i	D	Most nematodes are small soil organisms or parasites of animals or plants.
3.11i	C	This is a coral and belongs to the phylum Cnidaria.
3.12i	B	Brazil has a very large number of flowering plant species as it contains large areas of tropical forest (over 22,000 species).
3.13i	B	The Arthropoda has the most species. It includes all the insects and arachnids.
3.14i	C	Codes, for example the International Code of Zoological Nomenclature.
3.15i	D	A genus with only one species is monotypic; a family with only one genus is also monotypic.
3.16i	B	A pseudoextinction occurs when one line of a taxon evolves into a different taxon and all of the other lines die out. It therefore appears as if the original taxon has completely vanished.
3.17i	C	Wind carries pollen from male to female flowers in anemophilous plants.
3.18i	A	This describes Jordan's rule. The distracters are all fictitious.
3.19i	C	Jizz is correct. The distracters sound similar but are not biological terms.
3.20i	B	Cryptozoology is the study of evidence for the existence of species that have not been described by scientists.
3.1a	D	In a log-normal distribution only a very few species are common and many species are rare.
3.2a	D	The Wallace line is in Southeast Asia and runs through Indonesia, separating the Asian ecozone from the Australasian ecozone.

3.3a	C	Long before Linnaeus developed his binomial system of nomenclature Aristotle was putting organisms into natural groups that had particular characteristics in common.
3.4a	B	A conchologist is interested in the structure of shells. An oncologist is a cancer specialist; an entomologist studies insects; a mycologist studies fungi.
3.5a	C	$(2 \times 5)/(9+6) = 0.67$.
3.6a	B	A suborder lies below an order and above a family.
3.7a	D	Syntype is correct. A holotype is the single specimen that has been used to describe a species for the first time; a lectotype is a specimen retrospectively assigned to represent a species, replacing several syntypes; a genotype of an organism is the part of its genetic makeup that determines its characteristics.
3.8a	A	International Commission on Zoological Nomenclature is correct. The distracters are fictitious.
3.9a	B	In a trinomen there are three elements: genus, species, subspecies, e.g. *Panthera tigris altaica*.
3.10a	A	This rule requires that the oldest name must be used for a taxon.
3.11a	D	Some extinct animals are known only from signs of their activity such as footprints.
3.12a	A	A tribe is a taxon in zoology that occurs below family and subfamily but above genus.
3.13a	C	International Association for Plant Taxonomy is correct. The distracters are fictitious.
3.14a	D	Alexander von Humboldt was a naturalist and geographer who was an early pioneer of zoogeography.
3.15a	C	The answer is approximately 30 million $=(163/40) \times 100 \times 1.5 \times 50,000$.
3.16a	A	This is a species diversity gradient, with species numbers increasing from the Arctic (Alaska) to the Equator (Brazil).
3.17a	C	The horizontal axis $= \log_{10}$ of the area of the island; the vertical axis $= \log_{10}$ of the number of species. The relationship is described by a curve on linear scales, but taking the log of the values produces a straight line.
3.18a	D	In botany, division is generally used rather than phylum.
3.19a	D	Protists are animal-like, plant-like or fungus-like. They are eukaryotes, that is they have a distinct nucleus. Bacteria are prokaryotes and have no membrane-bound nucleus.
3.20a	A	Angiosperms are flowering plants and they are currently the dominant plant forms on Earth. Bryophytes are mosses, liverworts and hornworts; pteriodphytes are ferns, horsetails and lycophytes; gymnosperms are conifers, cycads and their relatives.

Chapter 4 Energy Flow and Production Ecology

4.1f	A	Green plants use light energy from the sun but some bacteria use energy from chemicals available in the environment.
4.2f	D	Charles Elton was a British ecologist who created the concept of a pyramid of numbers to study the relationship between the constituent species in a food chain or ecosystem.
4.3f	A	Each organism in a food chain breaks down carbohydrates and other biochemicals in the process of cellular respiration. This generates heat that is lost to the environment and this energy is not available to be passed to the next link in the food chain.
4.4f	B	A top predator or super carnivore eats other predators and is also known as an apex predator.
4.5f	D	Green plants are primary producers. The next link in a food chain is called a secondary producer or a primary consumer.
4.6f	C	The 'head' of a chlorophyll molecule has a magnesium ion at its centre which is essential to its function in capturing energy from light.
4.7f	D	The primary production of an oak woodland includes all of the plant material produced.
4.8f	C	A piscivore eats fishes.
4.9f	B	Not all heterotrophs are animals. Other organisms feed off plants and animals, including bacteria and fungi.
4.10f	A	Decomposers are bacteria and fungi involved in the process of decomposition.
4.11f	A	All of the organisms in the food chain lose heat when they respire and there are four points of energy loss between the phytoplankton and the killer whale not three.
4.12f	B	Chemotrophs obtain their energy from chemicals in the environment.
4.13f	B	T_1 = green plants; T_2 = herbivores; T_3 = carnivores; T_4 = top carnivores.
4.14f	C	A food web is a linear sequence showing simple feeding relationships, e.g. grass → rabbit → fox. A food web is a more complex representation of all (or at least the most important) feeding relationships in an ecosystem.
4.15f	A	Energy flows from one component of an ecosystem to another.
4.16f	B	Sanguivores feed on blood, e.g. vampire bats, mosquitoes.
4.17f	B	Some large fishes swim with their mouths wide open taking in food organisms as they move forward. This is called ram feeding.
4.18f	D	Some of the energy taken in from food is used in cell respiration. This arrow represents the heat loss as a result of this process.
4.19f	B	This is a coypu (*Myocastor coypus*). The presence of a diastema (a gap between the incisors and premolars) and the absence of canines indicate that this species is a herbivore.

4.20f	A	Oligophages eat few prey. From the Greek *oligos* (few) and *phagos* (eating from). Oligotrophic means having a deficiency of nutrients (e.g. oligotrophic lakes).
4.1i	D	Productivity is the rate of generation of biomass per unit area (e.g. $g/m^2/day$).
4.2i	D	A detritivore is an animal that feeds on dead organic material; decomposers (bacteria and fungi) break down organic matter.
4.3i	A	A is light from the sun, B represents green plants, C,D and E represent herbivores, carnivores and top carnivores respectively. G represents the heat energy lost in respiration. F is the decomposers that obtain energy from dead organisms and from animal wastes, etc.
4.4i	A	B is at the bottom of the pyramid and represents green plants (primary producers). A is the decomposers (bacteria and fungi) that feed on every other level but by convention they are represented by a column located on top of the primary producers.
4.5.i	C	This option describes a 'traditional' food chain in which the primary producers are more abundant than the herbivores and these are more abundant than the carnivores, producing a pyramid where the base is the largest section, unlike the diagram.
4.6i	C	Gross primary production is the energy trapped by green plants from the sun. Net primary production is what is left once the loss of energy from cell respiration has been subtracted from this.
4.7i	D	Consumption efficiency is the proportion of the energy in one trophic level (net productivity) that is taken in by the next level in the food chain.
4.8i	A	Some pyramids are widest at the top, e.g. when the primary producers are very large, such as trees, or when the food chain involves parasites.
4.9i	C	Tropical forests are the most productive biomes of those listed; they experience high temperatures, high light levels and high rainfall.
4.10i	D	This could represent a single tree being fed upon by many insect parasites each of which contains many endoparasites.
4.11i	B	This expression represents photosynthesis so C_6XO_6 is $C_6H_{12}O_6$ (glucose).
4.12i	C	An energy budget accounts for all of the energy that is taken in by an organism.
4.13i	A	Productivity should be measured in terms of dry biomass. The root systems are part of production but are difficult to measure and the relationship between the biomass of the aerial parts of the plant and the biomass of the root systems varies between species. If the relationship was consistent it would be a simple matter to exclude roots from calculations.

4.14i	B	Profitability is the return (in terms of energy intake) per unit effort (time) spent gathering food. Evolutionary theory would predict that animals would maximise this value.
4.15i	B	Gross assimilation efficiency = (food consumed – dung produced)/food consumed x 100. Dry weight values must be used so this becomes $((119 \times 0.3)-(139 \times 0.19))/119 \times 0.3 \times 100 = 26.0\%$.
4.16i	D	This is a process used by iron bacteria to obtain energy from ferrous carbonate.
4.17i	B	Elton noted that food chains were mostly restricted to five links. This is due to the energy lost at each stage.
4.18i	D	The digestibility of grass varies depending upon its chemistry, the time of year (new growth is more digestible) and the digestive system of the herbivore.
4.19i	A	Coprophagy comes from the Greek *kopros* meaning dung.
4.20i	B	Much plant material is made of cellulose and this is digested by cellulase. Some herbivores have a gut that contains bacteria that produce cellulase and aid digestion of plants.
4.1a	D	Photosynthesis takes simple chemicals from the environment and makes more complex compounds, thereby creating more order (less disorder) and thus causing a decrease in entropy.
4.2a	B	In complex food webs predators are able to switch to different prey species when their usual prey becomes scarce.
4.3a	A	The source of energy in a detritus food chain is dead organisms. In a grazing food chain energy passes from green plants to herbivores and then carnivores.
4.4a	D	The conversion of light energy to net primary production is inefficient in all ecosystems because light is reflected and lost in other ways, and plants use energy in cellular respiration.
4.5a	B	60% primary consumer (herbivore) because 60% of the food is plant material; 40% secondary consumer (carnivore) because 40% of the food comes from herbivores.
4.6a	D	Xylophagous is from the Greek root *xyl-* meaning wood.
4.7a	C	Lindeman's efficiency is a measure of ecological efficiency: the ratio of energy intake at successive trophic levels.
4.8a	C	This is the combined loss from heat, faeces and urine.
4.9a	A	As annual gross primary production increases so too does evapotranspiration.
4.10a	B	Only about 1% of the energy from the sun (insolation) ends up being used to produce more plant material by photosynthesis (net primary production) = NPP/insolation x 100.

4.11a	D	Carbon, nitrogen and sulphur are all important constituents of organic compounds and radioisotopes of these elements have been used to trace the movement of materials along food chains.
4.12a	B	Cell respiration breaks down glucose and releases energy in the form of adenosine triphosphate (ATP).
4.13a	D	Net primary productivity is the unit leaf rate multiplied by the leaf area index. NNP increases as the number of leaves increases and as the rate of production of biomass per leaf increases.
4.14a	B	((22/7) x 0.5 x 0.5 x23)/2 = 9.0.
4.15a	D	P-32 can be used to follow compounds containing P as they move along food chains; the precipitin test is a serological test that can be used to test gut contents for the presence of particular proteins from food organisms; visual analysis of gut contents can help determine what has been eaten.
4.16a	D	Small mammals grow (and therefore produce meat) more quickly than larger mammals from the same quantity of food.
4.17a	A	Eight hours is one third of a day so the gut contents represent a third of the food intake over 24 hours: 3 x 7.3 = 21.9. Dry weight is used so that the weight of water is excluded.
4.18a	C	As activity increases so does oxygen consumption because this is needed to release the energy from sugars. Species M uses slightly more energy at night, but species O uses more energy at night so is much more nocturnal.
4.19a	C	C4 plants (e.g. corn, sugarcane) are more efficient than C3 plants (e.g. rice, oats). C4 plants use the Hatch-Slack pathway for the dark reaction of photosynthesis and they are more efficient at fixing carbon dioxide than C3 plants. C2 and C5 plants are fictitious.
4.20a	A	Net growth efficiency is a measure of production per assimilation.

Chapter 5 Nutrient and Material Cycles

5.1f	A	This is a fungus and obtains its energy from dead and decaying organic material. From the Greek root sapr-meaning rotten.
5.2f	B	Pollution of water by organic wastes and fertilisers provides an excess of nutrients and causes excessive algal growth resulting in algal blooms.
5.3f	C	Nitrogen, potassium and phosphorus are the most important elements found in general fertilisers as they are essential for healthy plant growth.
5.4f	C	Leaching is the name of the process by which nutrients are washed from the upper layers of the soil by rainwater. Denitrification is a process in the nitrogen cycle.
5.5f	B	Guano is the name of the faeces produced by bats and seabirds and is used in some countries as a source of fertiliser.

5.6f	C	Calcium is important in the production of egg shells.
5.7f	D	Legumes include beans, peas and clover.
5.8f	A	Liming makes the soil less acidic and so makes nutrients more available to plants.
5.9f	C	This is the only process listed that is caused by humans so is the most recent.
5.10f	B	B is correct; the other options are the same values rearranged.
5.11f	D	The distracters are all fictitious.
5.12f	D	Nutrients come from physical and biological sources.
5.13f	B	Ammonification is a process that occurs as part of the nitrogen cycle.
5.14f	D	The rain is intercepted by the leaves and trunks of the trees with the result that water runs down the bark and drips off the leaves reducing the speed with which the rain reaches the ground and eventually flows into rivers.
5.15f	C	Beneath the water table the ground is saturated with water.
5.16f	B	Like other nutrients, some of the phosphorus in the fertiliser applied to the soil eventually washes out to the sea via the rivers.
5.17f	A	Mull is alkaline, porous, crumbly and decomposes rapidly, becoming well mixed with the mineral soil with no distinct layers apparent. It characteristically occurs in warm humid regions. Mor humus is acidic and occurs in soils where there are few organisms to mix the soil and decomposition is slower. It is characteristic of cold regions.
5.18f	B	This is a symbiotic relationship between a fungus and a plant (specifically the roots) and it plays an important role in plant nutrition.
5.19f	D	Nitrogen, oxygen and hydrogen are much more common in biological chemicals than sulphur.
5.20f	C	Dissolved salts in seawater consist mostly of chlorine and sodium ions.
5.1i	B	Evapotranspiration is the loss of water from plants (transpiration) plus evaporation from the ground and other surfaces.
5.2i	D	Hydrothermal vents are found in the deep oceans and also in some volcanic lakes. They enrich the water with nutrients that are utilised by organisms.
5.3i	B	Marine snow is made up of tiny particles of dead organic material that are slowly sinking to the bottom of the ocean.
5.4i	A	Plants absorb nitrate, nitrite and ammonium ions from the soil.
5.5i	D	Some fungi obtain their food by feeding as saprobes, some as symbionts and others as parasites.
5.6i	D	Catalytic soil enzymes occur in all three locations.
5.7i	A	These bacteria deplete soil fertility by breaking down nitrates and releasing nitrogen back into the air.

5.8i	A	Lightning fixes nitrogen in the atmosphere and causes the formation of nitrogen oxides.
5.9i	D	Tropical rainforest is dominated by very large trees and also contains many large animals. As a result some 80% of the organic carbon in these systems is in the organisms.
5.10i	C	The action of the wind on the surface of the ocean is to cause water movements that include upward movements from the deep ocean. These upward movements are called upwelling and are important in moving nutrients from depth to where the organisms exist in the surface layers.
5.11i	D	Zinc is only needed in very small quantities by plants.
5.12i	B	The main reservoir of carbon is the carbon dioxide in the atmosphere.
5.13i	B	Carbon dioxide in the air is used by green plants in photosynthesis to produce carbon compounds. These are consumed by herbivores and become animal carbon compounds. Cell respiration releases carbon dioxide to the air.
5.14i	A	Most of the water on the continents exists in the subsurface below the soil.
5.15i	B	Sedimentary rocks, for example limestone, contain most of the carbon found in rocks.
5.16i	B	Organic nitrogen is converted to ammonium by ammonification not assimilation. Assimilation is the process by which nitrogen is taken up from the soil and incorporated into plant cells.
5.17i	B	In this context advection is the flow of air horizontally over the surface of the sea.
5.18i	C	Water spends the longest time in the groundwater.
5.19i	B	*Nitrosomonas* converts ammonium to nitrite; *Nitrobacter* converts nitrite to nitrate.
5.20i	A	Some of the precipitation is unavailable to plants, for example because it evaporates from the soil before reaching the roots. The available precipitation is called the effective precipitation.
5.1a	C	The term allochthonous means formed, or originating from, elsewhere. Autochthonous means native to the place where it is found.
5.2a	A	Ammonium is converted to nitrite by the process of nitrification as the result of the action of autotrophic bacteria.
5.3a	D	Many forest floors are covered with a layer of seeds, acorns and other material that is collectively called mast.
5.4a	D	Sedimentary denitrification is an important nitrogen sink in seas and lakes.
5.5a	B	This occurs in the coldest zone.
5.6a	D	Flux rate should be measured as units/area or volume/day if rates are to be compared within and between ecosystems.

5.7a	D	After deforestation NO_3^-, Ca^{++}, K^+ and Mg^{++} were released and washed into the stream.
5.8a	A	The highest turnover rate is from the water mass to producers = (20/4)/(100/4) = 0.2. Note fluxes are calculated as units/acre/day so the values for fluxes and pools must be divided by 4 because the pond has an area of 4 acres.
5.9a	C	From the sixth year of application the total nitrogen in the soil is constant at 75 kg/ha.
5.10a	A	When iron is added there is a very large increase in primary production suggesting it is the limiting factor.
5.11a	B	Silicon is important in diatom growth.
5.12a	A	Nitrogen is present in only very low quantities in seawater.
5.13a	D	Unfavourable soil conditions, such as low pH, may make nutrients unavailable.
5.14a	C	The epilimnion is the surface layer and has few nutrients; the hypolimnion is at the bottom of the lake and anoxic.
5.15a	B	Nitrogenases fix atmospheric nitrogen into ammonia and are found in nitrogen-fixing bacteria.
5.16a	C	Trass is a type of volcanic ash; a wrasse is a type of fish.
5.17a	C	Evapotranspiration = precipitation – runoff – percolation. Distracter d is a rearrangement of this equation so is still true. Distracter b is accurate as evapotranspiration is the sum of losses from plants and the soil.
5.18a	B	The Haber-Bosch process is used in the industrial production of nitrogen fertiliser from the nitrogen in air.
5.19a	D	In section A the nutrient is limiting because as the quantity of the nutrient increases so too does the growth of the organism. In section B there is sufficient of the nutrient to maintain growth. In section C as the nutrient increases growth is suppressed as it reaches toxic levels.
5.20a	B	Oxidation would be fast if aeration is good, not slow.

Chapter 6 Ecophysiology

6.1f	B	Only mammals and birds can regulate their body temperatures physiologically.
6.2f	A	Leaves are shed by deciduous trees each winter. This protects cells from freezing temperatures.
6.3f	B	Allee's rule was devised by Joel Asaph Allen in 1877.
6.4f	D	A xerocole is adapted to environments where water is in short supply, e.g. a fennec fox, camel or addax.

6.5f	A	Homeostasis is the maintenance of a steady state (e.g. internal body temperature, the composition of the blood); homeothermy is the condition of being able to regulate temperature physiologically; homology is the similarity of features due to shared ancestry; homoplasy describes a feature that two or more species have in common that was not present in a common ancestor.
6.6f	B	Gigantothermy is the phenomenon whereby large ectotherms are able to retain heat due to their relatively small surface area. By definition such animals cannot be endothermic.
6.7f	D	Succulent plants store water in thick leaves, e.g. cacti, and live in environments where water is in short supply, e.g. in deserts.
6.8f	A	Stenoecious plants can only tolerate a narrow range of environmental conditions. Euryecious plants can tolerate a wide range of conditions. In plants, dioecious means having male and female sex organs in separate individuals. Monoecious means having male and female sex organs in the same individual.
6.9f	B	Birds use this method to lose heat in much the same way that some mammals lose heat by panting.
6.10f	A	Euryhaline species can tolerate wide changes in osmotic concentration in the environment. Stenohaline species can only tolerate a narrow range of salinity. An isohaline is a line on a map joining points of equal salinity. Hyperhaline refers to water about the salinity of normal sea water.
6.11f	D	The Loop of Henlé is the part of the kidney nephron that determines how much water is recovered during urine production. The longer the loop, the greater the quantity of water reabsorbed.
6.12f	A	Water is lost in the excretion of uric acid. By depositing uric acid around the body this water is saved.
6.13f	C	A hyperosmotic regulator maintains the osmotic concentration of its body fluids higher than that in the environment; a hypoosmotic regulator maintains it below that of the environment. Osmoconformers have a body fluid concentration that is the same as their immediate environment.
6.14f	A	Acclimatisation involves an animal altering the range over which a particular physiological variable is maintained. Animals can acclimatise, for example, to changes in temperature or oxygen concentrations in the air. Homeostasis is the physiological maintenance of a steady state in the body in relation to many variables, e.g. internal body temperature, blood chemistry.
6.15f	A	These hairs increase the surface area over which gases can be exchanged, e.g. in the hairy frog (*Astylosternus robustus*).
6.16f	D	Chromatophores is correct. A chromatocyte is a pigment cell; a chromatophyte is a type of alga.

6.17f	C	K. Schmidt-Nielsen has published widely on desert animals. The distracters are well known ecologists who do not have a special interest in these animals.
6.18f	D	Xerophytes live in a variety of habitats where freshwater is in short supply.
6.19f	A	High winds cause the stomata to close to reduce water loss.
6.20f	C	If the sun is in the north-west the locust would align its body south-west–north-east so that it is perpendicular to the rays of the sun and could maximise its heat gain. Locusts are poikilotherms so depend upon heat from the environment to maintain a body temperature capable of sustaining biological activity.
6.1i	B	Hibernation and aestivation are other types of dormancy.
6.2i	B	When the leaves are rolled up the stomata lose water vapour into the enclosed space. This increases the humidity immediately outside the stomata and so transpiration is reduced.
6.3i	A	Metabolic rate reduces during hibernation.
6.4i	B	Plants use carbon dioxide to make glucose in the process of photosynthesis.
6.5i	B	A tropism is a growth movement exhibited by plants. Aestivation is a prolonged torpor or dormancy exhibited by some animals to avoid a hot or dry period, e.g. some insects and amphibians.
6.6i	D	Salt is lost in urine and in the nasal glands near their eyes.
6.7i	A	Lizards regulate their temperature behaviourally by moving into the sun to warm up and into the shade to cool down.
6.8i	A	Plants use the relative length of day and night to determine when to flower as this changes with the seasons.
6.9i	D	The compensation point is the light intensity at which carbon dioxide uptake in photosynthesis is in balance with that produced by cell respiration. The other distracters are fictitious.
6.10i	C	Termites build large mounds which require ventilation so that gaseous exchange can take place.
6.11i	A	Frost drought occurs when soil water is frozen and therefore unavailable to plants.
6.12i	A	They survive largely on the water metabolised from seeds.
6.13i	C	Large bodies retain heat while small bodies lose heat because they have a larger surface area relative to their body volume.
6.14i	B	Named after the German zoologist Constantin Gloger. The distracters are fictitious rules using the name of well known biologists.

6.15i	C	This reflex allows the seal to remain submerged for long periods without the need to surface for air.
6.16i	B	When environmental conditions alter animal distributions change to reflect this, e.g. in response to climate change.
6.17i	B	This is a cactus with a fleshy body form and sharp spines.
6.18i	C	Phreatophytes are plants that derive their water from groundwater using a deep root system. They occur in deserts and semideserts.
6.19i	A	The distracters are fictitious.
6.20i	D	A refugee species is one that is confined to suboptimal habitats; an escapee species is one that is non-native and has established a wild population, e.g. after escaping from a zoo.
6.1a	A	This list combines the various methods different plant species use.
6.2a	B	Ectotherms need less oxygen than endotherms because the latter use oxygen to obtain energy from food, some of which is used to maintain their body temperatures. The internal temperature of ectotherms is determined by the external environment.
6.3a	B	Deer mice at a higher altitude would have a sigmoid dissociation curve to the left of that in the diagram. This indicates a greater affinity for oxygen at lower partial pressures such as would exist at high altitude.
6.4a	D	Camels can tolerate a wider range of body temperature than the other species. This allows them to cool down at night and start the day with a low body temperature which subsequently increases during the day.
6.5a	B	Chloride cells only occur in the gills.
6.6a	C	A counter-current heat exchanger transfers the heat generated by muscle contraction from the veins to the arteries in which blood flow is in the opposite direction. This retains heat in the body.
6.7a	A	Heliotherms use heat from the sun to warm their bodies; endotherms generate body heat internally.
6.8a	B	This equation describes the exchange of energy between an organism and the air by convection (air movement).
6.9a	A	Monocots have similar densities of stomata on upper and lower surfaces.
6.10a	D	This heat turns attractant chemicals into vapour that attracts pollinating insects.
6.11a	B	Subnivean animals live under snow, e.g. mice, shrews, voles and other small mammals.
6.12a	A	From the Greek *chion* which means 'snow' and *phile* which means 'lover'.
6.13a	D	Severe climate conditions in these areas depress tree growth and damage trees so that they take a stunted, twisted form.
6.14a	B	This means around the boreal region.

6.15a	C	Any increase in carbon dioxide concentration or light intensity would cause an increase in the rate of photosynthesis.
6.16a	C	After point C oxygen consumption increases rapidly and the bear's metabolic rate increases as it emerges from hibernation.
6.17a	D	Organic pollution decreases the availability of oxygen in water as it is used up by bacteria. Line D shows the best survival as oxygen level decreases.
6.18a	A	Red and blue wavelengths are used by chlorophyll; green wavelengths are reflected.
6.19a	A	Calcifuges live on acidic soils, for example *Erica* spp.
6.20a	B	At point B the net production of carbon dioxide is zero, i.e. the amount produced by cellular respiration = that used in photosynthesis.

Chapter 7 Population Ecology

7.1f	B	These animals cannot be a population because they are not all of the same species.
7.2f	B	Number of births per year.
7.3f	C	An irruption is correct. The distracters are simply terms with a similar meaning.
7.4f	A	An individual animal has a home range: the area it uses for its daily activities at a particular time. A population extends over a specific geographical area and this is called its range. A territory is a defended area within the home range of an individual. Habitat refers to a type of ecosystem, e.g. a savannah.
7.5f	D	Individuals possess a genotype, not populations.
7.6f	A	Plants per unit area = density.
7.7f	D	The size of a population is determined by its birth rate, death rate, immigration rate and emigration rate because these factors determine how many individuals are recruited into the population and how many leave.
7.8f	A	Frost will kill a proportion of a population regardless of its size so, although it will reduce the population it cannot regulate it. The other factors have a greater effect on larger populations than on smaller populations so have a regulatory function.
7.9f	A	The diagram is wide at the base, indicating a large number of young animals. As these grow up and reproduce they will cause an increase in the population.
7.10f	A	A group of animals of the same age is a cohort; they were born at the same time (e.g. in the same year).
7.11f	D	Crude death rate is deaths per unit time divided by population size (usually expressed as deaths per 1000); age-specific death rate measures the death rate in each age group; instantaneous death rate is the death rate over a very short period of time; finite death rate is the proportion of individuals alive at the beginning of a period of time that die during this period.

7.12f	B	A semelparous species reproduces once in its lifetime, e.g. some insects, octopuses and bony fishes.
7.13f	B	Pre-reproductive because she is too young to reproduce.
7.14f	C	The small groups of animals of the same species kept by zoos can be considered a metapopulation if they are exchanged for breeding purposes.
7.15f	D	By definition a population consists of individuals of the same species.
7.16f	A	This is a sink population because it absorbs individuals from elsewhere (a source population). A founder population is the original group of organisms that gave rise to an existing population.
7.17f	B	Demography is the study of populations: births, deaths, population structure, population growth, disease, etc.
7.18f	D	Infertile means incapable of reproduction; anoestrous means not exhibiting an oestrous cycle; impotent means inability to mate (in males).
7.19f	B	The term migration usually applies to a regular (usually annual) movement between two areas that are a considerable distance apart, often to locate a seasonally available food source.
7.20f	A	Lemmings exhibit population cycles which produce very large numbers and mass movements of animals.
7.1i	C	A morbidity table shows the proportion of individuals in each age group in a population that are expected to become ill. The other distracters are fictitious.
7.2i	D	The asymptote is where the curve has flattened and represents the carrying capacity of the environment at the time.
7.3i	B	The Lincoln index only works in closed populations. If marks are lost, this is the equivalent of marked animals emigrating from the population.
7.4i	D	This population is growing to the power of 2 with each time interval so it is exponential.
7.5i	A	Natality rate is the number of births per individual in the population per year.
7.6i	C	C is correct. The other options are simply the same names in a different order.
7.7i	C	The first occurs in spring (spring bloom) when temperatures and light intensity and duration increase. This then dies off releasing nutrients into the water. These nutrients fuel a second increase in autumn.
7.8i	B	In the absence of predators or competitors an introduced species – especially generalist feeders – may grow exponentially e.g. starlings (*Sturnus vulgaris*) introduced in the United States.
7.9i	A	A static life table contains the number of individuals in each age group at a particular time. A dynamic life table follows a cohort of individuals all born at the same time.

7.10i	D	B = l_x; C = e_x; a life table is concerned with deaths not births.
7.11i	C	Parasites produce many offspring, most of which fail to find hosts and die at an early age.
7.12i	A	The carrying capacity has fallen at this point to a new, lower level.
7.13i	A	This type of life table follows a cohort of individuals until they have all died.
7.14i	D	This is calculated as (20 x 12)/6 = 40.
7.15i	C	The classic interpretation of events here was that predators were removed from the area causing an increase in the deer and they subsequently overgrazed the area. A more recent analysis suggests that other factors were also at work.
7.16i	C	Age is plotted on the x-axis; The number remaining alive (survivors) is plotted on the y-axis.
7.17i	A	The description of a species as 'rare' has no quantitative meaning, unlike the distracters. A species can only be defined as rare when comparing its abundance with that of other (more common) species.
7.18i	B	The estimate would be calculated as (25 x 18)/10 = 45. If all ten marks had been retained in the population more should have appeared in the second sample, the number of recaptures would have been >10 so the estimate would have been lower.
7.19i	D	A life table shows either males or females because they are likely to have different survival rates.
7.20i	B	Some individuals may be caught once and never go near a trap again. Thereafter these are referred to as 'trap shy'. Other individuals may allow themselves to be trapped many times for the free food (bait). These are called 'trap happy'. Trap shy and trap happy individuals can affect the accuracy of population estimates.
7.1a	B	Named after Warder Clyde Allee. The distracters are fictitious.
7.2a	A	The variable r is the intrinsic rate of increase. The higher this value the faster the population will grow.
7.3a	B	The carrying capacity (asymptote) is denoted by K.
7.4a	B	The value of r is higher in population A as this is growing at a faster rate.
7.5a	A	The variable p_1 is the probability of an individual surviving from year 2 surviving to year 3.
7.6a	D	In the removal trapping method individuals are removed from the population and not returned. On each trapping occasion the size of the individual catch is plotted against the cumulative total catch up to that day. A line drawn through the points is used to predict the population size.
7.7a	C	This gallfly has a density-dependent effect on its host because a higher proportion of larvae were killed in 1935 (when the population of hosts was high) than in 1934 (when the population was low). It cannot be concluded from these data that all insect parasites have the same effect.

7.8a	A	If calving interval (the time between consecutive births) increases birth rate will decrease.
7.9a	A	Locusts have a high value for *r* and can increase in numbers quickly. The other species reproduce slowly.
7.10a	C	Elephants are *K*-strategists and reproduce slowly. The other organisms reproduce quickly.
7.11a	B	The product of the two sample sizes divided by the number of individuals they have in common. The proportion of marked animals in the second sample should be the same as the proportion in the population as a whole.
7.12a	A	Morbidity rate is a measure of the incidence of disease in the population.
7.13a	A	The oestrous cycles synchronise so that births occur together producing a surplus of easy prey for the predators and insuring that some individuals will survive.
7.14a	C	Environmental resistance is the combined effect of all the factors that prevent a population from growing exponentially.
7.15a	D	Fraser Darling was the ecologist who first described this effect. The distracters are fictitious effects.
7.16a	A	For example, if 10 plants produced 10 seeds each and 20% germinated there would be 20 germinating seeds. If there was a 10% chance of a seedling becoming established only 2 seedlings would survive. The total number of new plants produced is therefore 10 x 10 x 0.2 x 0.1 = 2.
7.17a	C	Only a process that has a density-dependent effect can *regulate* a population because it has a greater effect when a population is high (high density) than when it is low (low density). Density-independent processes have the same effect regardless of population density.
7.18a	B	This shows a population oscillating between an upper and lower limit. It neither grows uncontrollably nor becomes extinct.
7.19a	D	The graph shows the birth rate per individual in each age group.
7.20a	B	A stochastic model has an unpredictable outcome because it has a random element. Only model 3 is stochastic because it contains a random variable. Models 1 and 2 do not so are deterministic.

Chapter 8 Community Ecology and Species Interactions

8.1f	B	Zonation is a horizontal division within the habitat, e.g. on a seashore; succession is a temporal sequence of changes to a community.
8.2f	D	A vector or secondary host. A zoonosis is a disease that may be transmitted between an animal and a human.

8.3f	D	The term allopatric refers to a type of speciation; pioneer species are those that colonise new environments; megaherbivores are large herbivores such as elephants.
8.4f	B	Snowshoe hare and Canadian lynx.
8.5f	C	Cloud forest is a type of humid tropical forest.
8.6f	C	Taiga is boreal forest and occurs in North America and northern Eurasia.
8.7f	B	The term riparian relates to wetlands adjacent to streams and rivers.
8.8f	A	The benthos is the community of organisms living at the bottom of the seabed, a river, lake or other water body.
8.9f	A	The term 'ecto' means outer or external from the Greek *ektos*.
8.10f	A	Horizontal zones occur on rocky shores, largely defined by the organisms present. The distracters are names for other types of layering.
8.11f	D	Succession begins with a pioneer community and ends with a climax community.
8.12f	B	These organisms are lichens growing on the surface of a tree. The term 'epi' is from the Greek *epi-* meaning upon and *phyt-* meaning plant.
8.13f	A	A biome is a community of organisms in a major geographical region; a sere is a sequence of stages in the biological community during an ecological succession; an ecotype is a distinct form of a species occupying a particular habitat; an ecotone is a transitional area between two biomes.
8.14f	B	In a mutualistic relationship both organisms benefit. In the gut these microbes obtain food and shelter and in return they assist the utilisation of food that would otherwise provide little nutrition.
8.15f	A	Interspecific means between species. Intraspecific means within species.
8.16f	D	The distracters are all attributes of populations.
8.17f	B	A seral stage (sere) is a stage in the process of succession.
8.18f	A	The photograph is of a mangrove forest in Costa Rica. The trees here have a distinctive root system that extends above the mud.
8.19f	A	The dominant plant is marram grass (*Ammophila* spp.).
8.20f	D	The distractors are other types of unrelated aquatic systems.
8.1i	D	The pantanal is a large tropical wetland mostly located within Brazil.
8.2i	D	Insects in a tropical rainforest occupy a large number of narrow specialised niches; a tundra has a smaller number of wider niches.
8.3i	B	Species that occur in the same places can avoid competition with each other (interspecific competition) by occupying different niches (resource partitioning).
8.4i	C	The distracters are fictitious.

8.5i	B	Biological control is sometimes used in glasshouses to control agricultural pests.
8.6i	A	Species A is a prey organism and species B is its predator. Note that the peak in the prey numbers precedes that in the predators.
8.7i	C	An example of a guild is a group of unrelated taxa that use the same food resource.
8.8i	C	The relative abundance of different types of pollen grains allows palaeoecologists to reconstruct past plant communities.
8.9i	B	Niche breadth in tropical forests is narrow.
8.10i	A	Mountains are eroded with time so fossils rarely survive.
8.11i	A	Female *Anopheles* mosquitoes act as a vector for *Plasmodium* species that cause malaria.
8.12i	D	This is a secondary succession as a primary succession had previously occurred on this land before it was used for agriculture.
8.13i	D	Density-dependent mortality of plants, reduced seed production and a decrease in the production of vegetative offspring are all possible results of intraspecific competition.
8.14i	C	We would expect the diversity of insect species to increase with time as more niches become available.
8.15i	B	These species have dispersed from elsewhere, for example on the sea or carried by the wind.
8.16i	C	Boreal forest occurs at high latitudes in North America and northern Eurasia.
8.17i	D	In interspecific competition both species may be depressed (in numbers, growth, etc.) but they may evolve to coexist.
8.18i	A	Phytophysiognomy is concerned with the physical structure of plant communities, e.g. grassland, forest, etc.
8.19i	B	Savannah is at the base, followed by rainforest and then bamboo. At higher levels heath is replaced by alpine vegetation.
8.20i	B	Tropical savannah has a dry season and a rainy season and supports herds of large mammals such as zebra, wildebeest and elephant.
8.1a	C	References to environmental resilience usually refer to whole ecosystems and their ability to recover from perturbations.
8.2a	A	Paine studied predation in starfish. The distracters are fictitious studies.
8.3a	D	If niches are considered in three dimensions they will not overlap unless they overlap in terms of all three variables. If only two variables are considered they will overlap if they overlap in terms of just one variable.
8.4a	B	The species are using different parts of the trees for feeding for much of the time. This is an example of niche partitioning which helps them to avoid interspecific competition.

8.5a	D	When the ants and rodents were absent the seed density was 5.5 times higher than when both were present.
8.6a	B	Joseph Grinnell coined the term 'niche' in 1917. Haeckel was the first person to use the term 'ecology'.
8.7a	C	Hutchinson suggested that the niche of a species could be thought of as an n-dimensional hypervolume where each dimension represented a different environmental variable or resource.
8.8a	A	βN_1 alters the growth rate of species 2 based on the competitive effect of the size of species 1.
8.9a	C	Parasitism is considered to be a specialised type of predation as it is a type of consumer–resource interaction.
8.10a	B	A plagioclimax is one which cannot develop further due to continuous human activity, e.g. burning or grazing by domestic animals.
8.11a	D	This is an important book written by Robert MacArthur and Edward Wilson.
8.12a	C	Birds have colonised these islands from the mainland in the west. The islands encountered first by species dispersing from the west have the largest number of seed-eating bird species and the island furthest to the east has the smallest number.
8.13a	A	The Soviet biologist Georgy Gause performed laboratory experiments to test the Lotka-Volterra equations that describe the interactions of species on each other.
8.14a	B	MacArthur studied warblers in North American forest.
8.15a	D	This interaction is likened to an arms race because prey species evolve strategies to avoid their predators while the predators evolve strategies to catch their prey in a never ending race.
8.16a	A	When a predator's usual prey becomes scarce it will switch to a more common species. This avoids the predator wasting energy looking for animals that have become harder to find.
8.17a	B	Type I is a linear response; type II is correct; type III is similar to type II but is sigmoidal in shape; type IV does not exist.
8.18a	A	Predators seek to maximise energy intake and minimise energy costs so may switch diets to achieve this.
8.19a	C	The total number of potential routes from primary producers to the top carnivores or if there are none, the highest consumer level.
8.20a	D	*Fasciola hepatica* is the sheep liver fluke, an economically important platyhelminth parasite with a worldwide distribution, that affects a range of mammals, including humans, causing fascioliasis.

Chapter 9 Ecological Genetics and Evolution

9.1f	D	If mutation did not occur the only source of variation within populations would be that resulting from the recombination of genes. Natural selection and genetic drift influence which genes and alleles survive in a population but do not themselves create new variation.
9.2f	B	A grazing habit has evolved separately in kangaroos in Australia and antelopes in Africa. These taxa are not closely related.
9.3f	C	Sympatric speciation is the evolution of new species from an ancestral species while their distributions continue to overlap.
9.4f	B	In Müllerian mimicry two or more species that have evolved defences such as an unpleasant taste, and that share common predators, mimic each other to their mutual benefit. In Browerian mimicry (automimicry) some (defenceless) members of a species gain an advantage by their resemblance to other members of the same species (with defences).
9.5f	A	Wallace and Darwin came up with similar theories of evolution at more or less the same time.
9.6f	C	By continually breeding from drought-resistant individuals, artificial selection has produced drought-resistant crop varieties.
9.7f	B	Evolution occurs when the natural variation in biological populations results in some individuals surviving and reproducing while others (with different genes) die and fail to reproduce. This variation is the raw material of natural selection and without it evolution could not occur.
9.8f	C	Adaptive radiation is the diversification of life forms from a common ancestor so that they are able to fill different ecological niches.
9.9f	A	A cline is a gradual change in one or more characteristics often between different populations; a clone is an individual that is genetically identical to another; a meme is an element of culture passed from one individual to another by non-genetic means.
9.10f	B	Genetic load is the reduced fitness in a population compared with the fitness the population would have if it consisted solely of individuals with the reference high-fitness genotype.
9.11f	D	Mutation provides new alleles, recombination mixes alleles up between generations and selection determines the frequencies of alleles in a population.
9.12f	A	A small amount of gene flow will allow populations to evolve independently because they are almost completely isolated from each other.
9.13f	B	Natural selection favours individuals that are well adapted to the prevailing environment.
9.14f	C	Inbreeding is breeding between close relatives, interbreeding generally refers to breeding between different species (hybridisation), and speciation is the creation of new species during the course of evolution.

9.15f	A	Microevolution is correct. Macroevolution is evolution at the level of species or higher.
9.16f	A	*The Origin* was published in 1859.
9.17f	D	Adaptation is the process by which a species becomes better suited to its environment.
9.18f	B	Fitness relates to reproductive success and the contribution made to the gene pool of the next generation.
9.19f	D	Genomics is a relatively new multidisciplinary area of biology concerned with the structure, function, mapping and editing of genomes, and with evolution.
9.20f	B	Ecological genetics is the branch of genetics that is concerned with natural field populations as opposed to populations of organisms kept in the laboratory (classical genetics).
9.1i	B	Genetic drift is the change in the frequency of an allele as the result of random processes.
9.2i	A	Industrial melanism is the prevalence of dark coloured varieties of organisms in industrial areas where they are better camouflaged against structures such as dirty buildings than paler variants.
9.3i	B	Birds with smaller bodies were less likely to survive in cold weather than those with larger bodies as the latter were better able to retain heat. Body size is controlled by several genes and the increase in the frequency of alleles for larger bodies would result in an increase in the mean body size in the next generation. This is an example of directional selection.
9.4i	C	Allopatric speciation is caused by geographical isolation; parapatric speciation occurs when two populations of a species that are reproductively isolated from one another continue to exchange genes; convergent speciation is fictitious.
9.5i	C	Inbreeding results in the accumulation of deleterious alleles and reduced fitness.
9.6i	A	The effective population size is often smaller than the actual population size because some individuals do not breed or the sex ratio is not 1:1.
9.7i	A	Metal-tolerant species of some plants have evolved due to exposure to metals in the natural environment, e.g. grasses growing on mine waste. In such environments natural selection favours tolerant genotypes.
9.8i	D	Genealogy is the study of family history; phylogeny is the study of the evolutionary history of organisms; phylogenealogy is the study of the lines of descent of groups of males.
9.9i	C	He has 50% of the genes of each of his parents and they have 50% of the genes of each of their parents, so the baboon has 25% of his grandfather's genes.
9.10i	C	Kin recognition presumably evolved to prevent inbreeding.

9.11i	D	All of these terms mean the same thing. In Scotland the gene pool of the European wild cat has experienced genetic pollution as a result of interbreeding with domestic cats.
9.12i	B	A molecular marker is a fragment of DNA associated with a specific region of the genome.
9.13i	D	Sexual selection is a mode of natural selection which has favoured the evolution of certain conspicuous traits because they give the possessor of such traits greater access to members of the opposite sex and thus confer greater fitness.
9.14i	A	The expression of the phenotype is a result of the interaction between the genome and the environment.
9.15i	D	Isolation of this lion population reduces mate choice and thus increases the probability of mating with a close relative.
9.16i	B	The distracters contain the names of other famous geneticists.
9.17i	A	Kettlewell conducted field experiments on *Biston betularia* in industrial areas of England.
9.18i	B	Hybridisation has occurred between polar bears (*Ursus maritimus*) and brown bears (*U. arctos*) possibly as a result of the latter species extending its range.
9.19i	B	Animals can increase their genetic fitness by reproducing themselves or by helping relatives to produce offspring, because they have some of the same genes.
9.20i	D	PCR is a method used to amplify small quantities of DNA so that it may be analysed.
9.1a	D	Although Darwin discovered these species the detailed work on their adaptive radiation was conducted much later by David Lack who published a book entitled *Darwin's Finches* in 1947.
9.2a	A	Inbreeding occurs when close relatives mate. When mating occurs between distant relatives outbreeding depression occurs. Both situations can result in reduced genetic fitness.
9.3a	C	The effective population size will be less than the actual population size because the sex ratio is biased in favour of females. The actual value is $(4 \times 109 \times 141)/(109+141) = 246$.
9.4a	D	Genetic diversity will have been reduced because of a sudden reduction in the gene pool when the bottleneck occurred.
9.5a	B	A cline is a gradual change in morphology (e.g. colour) extending across the range of a species.
9.6a	C	This diagram is specifically called a cladogram based on concepts used in cladistics.
9.7a	C	Ford published widely on the ecological genetics of moths and butterflies.

9.8a	D	This is a ring species. The other terms are fictitious.
9.9a	A	The shell colour is the result of natural selection.
9.10a	B	The diagram shows the relationship between gene frequencies and genotype frequencies in the Hardy-Weinberg equilibrium.
9.11a	D	The hybrid zone is approximately 20-50km wide. The carrion crow occurs in western Europe and the hooded crow occurs in eastern Europe.
9.12a	B	This is a filter route. A sweepstake route is one where dispersal occurs largely by chance because surrounding environments are largely unsuitable; a corridor is the easiest type of dispersal pathway where the chances are good that most species will spread.
9.13a	C	A loss of genetic variation occurs when a new population is established from a very small number of individuals. This is the founder principle or effect.
9.14a	D	If an animal helps an unrelated individual this is unlikely to result in a fitness payoff for the helper because they do not share similar genomes.
9.15a	C	In population C each individual possesses two recessive alleles (dd) and the dominant allele (D) is absent.
9.16a	B	$p + q = 1.0$. $q = 0.3$, $p = 0.7$. The frequency of Aa $= 2pq = 2 \times 0.3 \times 0.7 = 0.42$.
9.17a	B	A ring species is illustrated in Fig. 9.5.
9.18a	C	For the Hardy-Weinberg equilibrium to apply mating must be random.
9.19a	D	The frequency of ff is $q^2 = 0.28$. $q = 0.53$; $p + q = 1.0$, so $p = 0.47$. The frequency of FF $= p^2 = 0.22$.
9.20a	A	An ESS becomes fixed in a population as a result of natural selection.

Chapter 10 Ecological Methods and Statistics

10.1f	B	A Baermann funnel is designed to extract organisms that live in soil water films. Could be confused with a Tullgren funnel.
10.2f	A	This is a device used for weighing objects by attaching them to a hook fixed to a spring. As the spring stretches a pointer moves down a scale indicating the weight.
10.3f	C	A bird observatory may be permanently occupied by one or more bird ringers who record passing birds that may be trapped in Heligoland traps.
10.4f	B	Autonomous submersible vehicles are essentially robot submarines capable of capturing video, measuring environmental variables and collecting samples, often operated from a research ship.

10.5f	A	A Sherman trap is a live trap for small mammals that is made of metal and hinged so that it folds flat.
10.6f	C	This is an insect trap that essentially consists of a light source enclosed within a fine mesh curtain.
10.7f	B	Grapnels are used to grab samples of macroscopic aquatic plants.
10.8f	B	A vasculum is a metal case used by botanists to protect plants collected in the field from being crushed. It typically takes the form of a flattened cylinder with a hinged door on one side and a shoulder strap.
10.9f	C	A beating tray is a large flat receptacle that is held under the tree or bush while the branches are struck.
10.10f	A	Citizen science projects involve participation by large numbers of members of the public in the collection of data, for example bird or butterfly surveys. The data are then analysed by scientists.
10.11f	C	Plants touched by the metal pins in the frame are included in the sample.
10.12f	D	Harp traps are used to catch bats in flight without the risk of them becoming entangled in a net. They fly into strings suspended from a frame and fall into a collecting chamber.
10.13f	B	A fine insecticide released from a blower is used to knock insects from trees and bushes.
10.14f	C	A seine net is used to catch fishes. It may be towed behind a boat at sea or held by fishermen across a river.
10.15f	A	A Longworth trap is a live trap for small mammals.
10.16f	B	A mist nest has a fine mesh and is suspended between trees or other structures to catch small birds.
10.17f	A	A pooter is a small transparent glass or plastic vessel used to collect small insects. Air is sucked out of the vessel through a tube and the insect is drawn into the container through a second tube.
10.18f	B	A pitfall trap is a cup-shaped, smooth-sided container that is placed in the soil so that its top edge is level with the soil surface. The container is usually covered with a stone suspended slightly above the top to disguise it and prevent it filling with water.
10.19f	D	Variance and standard deviation are measures of dispersion around the mean.
10.20f	C	A continuous variable can take on any value between its minimum and its maximum value. Colour is a discrete variable because it can only take on a limited number of values.
10.1i	B	A Tullgren funnel consists of a funnel over which a soil or leaf litter sample is suspended on a mesh. A light is suspended above the sample and small soil organisms, mostly arthropods, fall through the funnel into a bottle containing alcohol.

10.2i	B	Bomb calorimetry involves burning samples and measures the heat of combustion in a chemical reaction.
10.3i	A	This is a widely used technique used to monitor water quality by invertebrate sampling.
10.4i	D	The field of view is largely determined by the focal length and the magnification. Binoculars with a x8 magnification have a wider field of view than those with a x10 magnification.
10.5i	D	These are large, heavy binoculars which would be cumbersome in the field and the high magnification would reduce the field of view making it difficult to locate small birds and difficult to hold steady.
10.6i	C	A transect must follow a line. The quadrats in C were thrown at random.
10.7i	C	A Chi-squared test would allow you to determine whether or not 175 males and 197 females is significantly different from 186 of each sex.
10.8i	C	Although quadrats are usually square the term is sometimes used for other shapes of sampled areas, for example rectangular, circular or irregular in shape.
10.9i	B	In a normal distribution the curve is bell-shaped and the mean, mode and median all occur in the middle.
10.10i	D	This is a clumped distribution with most samples containing few or no plants.
10.11i	A	Environmental DNA (eDNA) is collected from the environment (e.g. a stream) and its composition allows the presence of rare species to be detected that might otherwise be overlooked.
10.12i	D	The mean density of plants was 16.76 per m^2. The population is therefore 16.75 x 55 = 921.
10.13i	D	The soil core provided evidence of a temporal sequence (succession) that began with the taxa found at the bottom (reeds).
10.14i	D	Only the census taken in 1990 and 2000 can be directly compared because the others used different methods. From this evidence there appears to have been an increase in numbers from 1990 to 2000.
10.15i	A	This method requires the researcher to close up entrances to burrows and return later to determine which have been reopened. Those burrows that have not been reopened are assumed to be unoccupied.
10.16i	D	A simulation uses a computer programmed with mathematical equations that represent some process in the environment, e.g. the effect of predator numbers on their prey, to predict outcomes when the values of certain variables are changed.
10.17i	B	Historically, most biologists have not been trained in mathematics. For this reason some of the early mathematical models were devised by individuals with backgrounds in physics, engineering and related disciplines.

10.18i	C	The logistic equation is one of a number of mathematical models used to describe and investigate population growth.
10.19i	C	A bathyscope is a device similar to a bucket with a transparent bottom that is placed against the surface of a body of water to reduce reflections and improve underwater visibility.
10.20i	A	The use of the median (the middle value when values are ranked) allows exceptionally high and exceptionally low values to be ignored.
10.1a	C	A correlation coefficient can range from +1 (perfect positive correlation) to -1 (perfect negative correlation). A value of zero indicates no correlation.
10.2a	A	A type I error has occurred when a statistical test suggests an effect has been observed when the result is actually inconclusive (false positive).
10.3a	C	The GPS unit gives the location of the observer. If we know the bearing to the tiger and its distance from the GPS we can calculate the tiger's location using trigonometry.
10.4a	A	There are no data provided on traffic volume and any comparison between months assumes traffic volume stays constant and the number killed is directly proportional to population size.
10.5a	D	The values of these two indices cannot be compared because they use different assumptions. To make valid comparisons the same method must be used at both sites.
10.6a	D	An independent t-test would determine whether or not the two samples had been taken from the same population, i.e. whether or not there was a significant difference in their means.
10.7a	B	Regression analysis plots the dependent variable (plant growth) against the independent variable and calculates the position of a line through the points.
10.8a	D	Stochastic models take account of random events so they produce a different outcome each time they are run.
10.9a	C	The lowest evenness is seen in the community where the individuals are least evenly spread between the species. In community C most of the individuals belong to species M.
10.10a	C	Between 1 and 5 samples the population estimate was changing considerably with each additional sample. After 12 samples had been taken the estimate remained quite stable so any additional sampling did not substantially alter it.
10.11a	A	Nearest neighbour analysis requires the distances between individuals to be measured. It does not require a clearly defined sample such as that obtained using a quadrat.

10.12a	B	This distribution shows highly clumped individuals with few individuals between the clumps. If this type of distribution was sampled using a quadrat the variance of the number of plants counted in each quadrat would be greater than the mean so the value of I would be >1.
10.13a	C	A Monte Carlo simulation has a stochastic (random) element so the outcome would be different each time it was run.
10.14a	A	A 2x2 contingency table should be used to calculate a Chi-squared value.
10.15a	D	The different mesh sizes restrict the types of decomposers that can enter and begin the breakdown of the leaf litter.
10.16a	A	A McMaster counting slide is used for counting parasitic worm eggs in faeces. The distracters are fictitious.
10.17a	B	The mean number of animals per transect = 6. Transect area = 700 x 150 = 105,000m². Population estimate = 1,600,000/105,000 x 6 = 91.4.
10.18a	D	Acoustic lures draw animals such as bats and birds closer to observers so that they may be counted.
10.19a	C	A kite diagram is a graphical representation that may be used to map the abundance of plants along a transect.
10.20a	B	A cluster analysis can be used to produce a diagram representing the relationship between different samples (in this case soils) based on similarities in a number of variables. Those samples linked by horizontal lines at the lowest threshold level (on the vertical access) are the most similar to each other.

References

Cain, A. J. (1971). *Animal Species and their Evolution.* Hutchinson & Co. (Publishers) Ltd., London.

Collier, B. D., Cox, G. W., Johnson, A. W. and Miller, P. C. (1974) *Dynamic Ecology.* Prentice/Hall International Inc., London.

Johnston, D.W. and Odum, E.P. (1956) Breeding bird populations in relation to plant succession on the Piedmont of Georgia. *Ecology* 37, 50–62.

Laws, R. M., Parker, I.S.C. and Johnstone, R. C. B. (1975) *Elephants and Their Habitats. The Ecology of Elephants in North Bunyoro, Uganda.* Oxford University Press, London.

Menzel, D.W. and Ryther, J.H. (1961) Nutrients limiting the production of phytoplankton in the Sargasso Sea, with special reference to iron. *Deep Sea Research* 7, 276–281.